"... No one on Earth has ever seen you."

"Mr. Starkman, I am sitting less than ten meters from you at this very moment, watching you through one-way glass. It is true that I have not permitted myself to be seen by more than a handful of your species; this is because of the strong tendency toward xenophobia that has been manifest throughout human history. I am very alien indeed, a thing fit to appear in one of your horror movies..."

THE CHROMOSOMAL CODE

LAWRENCE WATT-EVANS

AVON
PUBLISHERS OF BARD, CAMELOT, DISCUS AND FLARE BOOKS

AVON BOOKS
A division of
The Hearst Corporation
1790 Broadway
New York, New York 10019

Copyright © 1984 by Lawrence Watt Evans
Published by arrangement with the author
Library of Congress Catalog Card Number: 84-90800
ISBN: 0-380-87205-6

First Avon Printing, May, 1984

Printed in the U. S. A.

WFH 10 9 8 7 6 5 4 3 2 1

Dedicated to
PHILIP K. DICK
whom I never met
but who was long
my favorite author.

Chapter One

The landscape was white and gray and black, with no trace of color. The sky was the dull, dead, dirty off-white of winter on this June day; the ground was hidden beneath the pale white sheen of clean snow. The building ahead of him, what little of it showed, was the rough white of painted concrete, picked out with the gray and black of shadows. Black shadows also served to mark the pine trees that edged the open area that had once been a parking lot; the dull green of their remaining needles was lost in the sodden masses of snow that weighed them down.

The building's façade was largely hidden in a long strip of shadow where the overhang failed to meet the snowdrifts, a strip perhaps fifteen meters long and one meter high, divided into segments by the peeling white-painted, unornamented pillars. At one point to the right of center the lower edge of the gap dipped down, the sharp white edge of the drifts broken through. His dirty brown coat the only color in sight, John Starkman was headed toward this break in the wall of snow.

The footsteps he followed through the drifts were his own, layered on top of one another but all of his own making, the only break in the smooth surface of the snow and the only trace of human presence. He had discovered the building a month or so ago, astonishingly, miraculously intact, and had made the long trek out to it half a dozen times. It was such a rich find that he had seriously considered moving from his long-established home to somewhere nearer at hand.

On a day like this, however, he didn't mind the walk.

Enjoying the unaccustomed June warmth, with temperatures as high as three or even five degrees above freezing, Starkman had the fur-lined hood of his coat flung back; the gentle breeze, so unlike the freezing gales of winter, kept his untidy hair clear of his eyes. There was little glare under the leaden, overcast sky, and no chance that he could imagine of encountering another human, so he had left his sunglasses in his pocket.

He reached the base of the final barrier of snow and clambered up the drift, his feet packing down the powder that had blown into his path since his last visit. For a moment he was reminded of his New England childhood, when climbing drifts had been something he did for fun, rather than out of necessity.

He half climbed and half slid his way down the other side, catching himself upright again on the narrow strip of sidewalk left bare by the encroaching snow. Before him in the shadowed dimness was the man-sized hole he had broken in the glass front of the little supermarket; the only trace of snow beyond the glass floated in a puddle of dirty water just below the opening, where it had either been blown by a freak of wind or been tracked in on Starkman's boots. He reminded himself that he would have to clear that puddle away somehow if he was to avoid having an inconvenient patch of ice when the warm spell ended. He stepped carefully through the hole and strode past the checkouts into the darkness of the deserted aisles, aisles that were still stacked high with unlooted merchandise. The sound of his bootheels on tile turned sharp and clear as the slush and snow fell away.

He marveled anew that even here, so far from the city, no desperate fleeing party or lingering scavenger before him had come across the place and stripped it clean. It was well off the main roads south, and in an area where the neighbors would have been rich enough to leave early, before looting became common, but it was still surprising.

He found what he wanted and emerged into the dim light at the front of the store with a bottle of Coke in hand. He seated himself comfortably on one of the checkout counters and reached down beside the cash register to retrieve the bottle opener he kept stashed there. When the cap was pried off the bottle fizzed gratifyingly, which he considered a fairly reliable sign that the stuff was still safe to drink. He looked at it critically, noting that there was only a slight trace of ice, then took a long pull at the bottle before resting it once more on his knee and gazing out at the broken window and the barrier of snow beyond.

It was his custom to take a rest like this before getting down to the serious business of gathering supplies and carrying them back home; the cola was a regular part of it, since the journey always made him thirsty and a little weary. He was convinced that the sugar helped provide the energy necessary for the return trip.

Sometimes, if there was sufficient light, he would read the paperbacks and the browning tabloids that were ranged around each checkout, challenging himself to identify each of the celebrities the scandal sheets mentioned. On this occasion he just sat, staring out at the snow and taking a swig from the bottle every few seconds, admiring the plate-glass frontage. Except for the entry hole he had made himself, it was still completely sound, as though the supermarket might open for business at any moment. Faded signs advertising specials on chicken and lettuce still clung to the glass, while others had fallen to the floor in mild disarray. One poster neatly covered the glass he had swept aside after breaking his entrance. The fluorescent lights, both in the store and under the overhang out front, were unbroken, as if the flip of a switch could bring them to life as it had long ago.

The drifts beyond were not more than three meters high at the very most, probably closer to two and a half, and easily a dozen centimeters lower than on his last visit. He wondered how long the good weather would last, and whether he might see bare ground without having to dig for it. That hadn't happened for ten years.

Such a complete thaw seemed unlikely. Probably, he told himself, the warm spell would pass, and the summer would be neither better nor worse than those that had preceded it.

He took another long pull on the bottle, then put it down. The puddle under the window caught his attention, and he tried to remember which aisle held buckets and mops and sponges.

Something moved at the edge of his vision; he started and looked up.

Something was moving out there, in the area of sky visible between the snowdrifts and the overhang, a dot against the gray-white overcast. His eyes widened from their customary snow-blind squint.

He tried to tell himself that the dark speck was a bird, some damn fool of a bird that had found its way north again, but the thing was not moving like a bird, with swoops and flutters or riding the breeze. Instead it moved steadily and slowly through the air, along a dead-straight course. It glinted dully despite the clouds.

Starkman sat still for a long moment, too startled to move. When it finally registered that the thing was not only moving but coming closer, he got quickly to his feet and stepped out through the broken window.

The first sound reached him, a low throbbing rumble, and his surprise increased beyond what he had thought possible. What he had taken for a plane—and a plane would be quite astonishing enough after all this time—was something else. The sound was wrong. As it drew nearer he saw that the shape was likewise strange. It was formed something like an Indian arrowhead he had had as a boy, a convex-edged triangle forming the prow and a constricted waist joining it to a boxy stern assembly.

There could be no doubt whatsoever that it was a machine, and a machine like nothing he was familiar with. There were no wings, no rotors, no visible jets, no contrail, yet it was flying. It was gliding smoothly almost directly toward him, its upper side gleaming metal, the details of the rest lost in shadow.

It seemed to approach for an impossibly long time; it was vastly larger than he had realized at first, larger than he could judge with any accuracy. There were no visible features to give him a scale; the thing's surface appeared smooth and unbroken. He could see no doors, hatches, windows, ports, antennae, or anything else except gray metal.

The rumbling sound grew steadily until it was a deafening roar as the machine passed directly overhead. It seemed to clear the overhang by mere centimeters, a great dark expanse sliding along above the supermarket no faster than a man could walk, blocking out the feeble filtered sunlight and leaving him in the awesomely complete darkness of its shadow. The vibration of its passing shook the building, rattling cans and bottles within, and somewhere among the aisles a bottle toppled from its shelf with a crash barely discernible over the battering thunder.

Even at its closest approach Starkman could see no detail on its surface, no welded seams, no rivets, no lights of any sort, no openings or vents. Of course, he knew that the shadows might be hiding them. He was certain that the machine must be farther away than it appeared, though it was definitely at a very low altitude. That meant that it was even larger than he had thought. He stared up at the featureless metal.

The head of the thing was past, and the darkness receded slightly as the narrow waist moved over, light seeping around its edges. It was as featureless as the prow. The tail assembly restored the deep shadow once again, but he thought he glimpsed irregularities in its surface, though he could not decide what they might be, since they were indistinct in the gloom.

As the trailing edge of the immense machine moved on past the overhang and out of sight, he remained motionless, staring up at the empty sky it had left behind and blinking in the renewed light.

The noise receded slowly, somewhere behind him, for a long moment; then, while still a roar worthy of a good-sized waterfall, the sound stopped declining and held steady. Starkman whirled, as if expecting to see through the concrete rear wall of the supermarket.

The sound changed, suddenly swooping higher in pitch and altering its tone until it was more a whistle or a scream than a rumble, though still an earth-shaking bass. A hissing blended in an unearthly cacophony, growing in volume until, abruptly, the sound died, as if the plug had been pulled on some immense factory machine.

Starkman remained motionless for a long moment of the new silence, a silence as complete as before he had first seen the thing approaching. Then, with sudden decisiveness, he turned and scrambled up the snow-drifts. From the top of the drift beside his broken-through path he could easily reach the edge of the overhang; he grabbed the concrete, digging the tips of his fingers into the snow it supported, and dragged himself up, pushing up onto his elbows and then crawling forward into the snow.

The supermarket's sign lay under the snow, long since fallen back, its plastic face buckled under the weight of the snow; he landed atop it and felt it break beneath his arms. He heard the fluorescent tubes within the sign splintering. He backed off, wary of cutting himself, and pulled himself sideways along the edge of the roof until he was past the end of the old display. Then he hauled himself completely up onto the roof and stood, cautiously.

He could see nothing but snow; the drifts on the unheated roof were very nearly as tall as those on the ground.

Annoyed, he climbed up a drift; the partial thaw had left the snow wet and dense, for the most part, so that he had little trouble with slippage. When his head was above the top of the white mass he leaned forward and peered over.

The ship—he could not help thinking of it as that— had landed. It lay quietly on the vast expanse of open, untrodden snow that had once been a golf course, plainly visible between the tops of the pine trees that formed a row along the back of the store.

Now that it was grounded, with trees providing reference, Starkman was utterly appalled by the size of the thing. It was easily a hundred meters wide, at least twice that in length, and half as much, perhaps, in height. It had landed at a slight angle to him, so that he could see the entire stern and part of the left side. He had a good view of what were surely immense exhaust ports, though he had seen no evidence at all of any exhaust while the thing was in flight. He remembered movies he had seen as a child and books he had read since, and had no doubt what he was looking at.

This, he was convinced, had to be a starship.

He backed down the snow toward the edge of the roof, then lowered himself down to the drifts below and let himself slide down toward the store windows. He lay there on his back for a long moment, staring unseeing at the announcement of "WHOLE FRYERS $1.89 KG." as he let it sink in what he had just seen.

Damn, he thought, a starship; people from another star system. Why hadn't they come twenty years ago, before the weather changed and the snows came?

The first doubts began to seep in. How could a starship be here? Wouldn't they have landed somewhere warmer, where there were still signs of civilization? He was sure that there must be civilization somewhere in the tropics. And what were the odds that he, himself, would just happen to be the only human present at the landing of the first starship to reach Earth?

Ignoring the sudden guess that maybe it wasn't the first, he decided that he must consider all the possibilities, one of which was that this thing was merely a new sort of atmospheric craft—or even a spaceship—built in some southern country untouched by the ice age. The more he thought about it the more he convinced himself that that must be the case; they were undoubtedly searching the northern countries for survivors such as himself, offering to take them to South America or the South Pacific or wherever it was still warm.

The very resemblance of the ship to starships in old movies suddenly struck him as strong evidence against its extraterrestrial origin; it was just too familiar, too human in its design.

But its size—why would a ship searching for survivors be so huge?

Other theories crowded into his mind; perhaps the ship was a self-contained community, landed to replenish its stocks of water and raw materials. Perhaps it was a warship taking a circuitous route from one combatant in a tropical war to the other, hoping to avoid detection; the smooth surface might be reflective armor against laser weapons or radar detection. Perhaps it was a shipload of undesirables being exiled to the frozen wastes.

And in any of these cases, what should he do about it? Should he go to greet the new arrivals, or flee for his life?

He resolved, after some thought, on a compromise; he would scout out the ship, as cautiously as he could, and keep a careful watch on it. First and foremost, he wanted to see what its crew looked like. Despite the misconceptions of science-fiction writers, he was quite certain that alien beings would necessarily look *alien*. He was convinced that the theory of parallel evolution was garbage; he had studied genetics in school, and knew that the odds against the same long series of accidents of evolution occurring twice were astronomical, even if the planet in question had conditions identical to Earth in all respects—a very dubious proposition in itself. If the crew of the starship—or rather, of the ship—looked human, then they *were* human, and merely from some surprisingly advanced culture in the south, not from another star.

If they were aliens—the idea seemed much less likely with every passing second—he would flee, probably heading south to try to contact some sort of authority equipped to deal with such a thing. At any rate, some sort of authority was wanted, though he doubted that anyone would be equipped to handle something like first contact with aliens. He was not about to take such an immense responsibility and risk upon himself.

If they were humans he would watch them closely, assess their behavior, and decide what to do when he had some idea of who and what they were.

This decided, he got to his feet, turned, and started up the drift again. A pace short of the top he froze, looking out across the snow of the empty parking lot.

His footprints, overlaid upon each other, made a clear, straight mark pointing him out to anyone who cared to look.

He hadn't worried about enemies in years; he hadn't needed to. The last lingering holdouts besides himself had given up almost a decade ago, after the second year when the snow never melted in the summer. For a year or two after that he had feared that others might turn up, coming from cities farther north or spreading out from towns where supplies had been exhausted. They hadn't come. For a year or two after that, when the solitude became oppressive, he had hoped that others would come; they still hadn't. Finally he had simply forgotten the whole subject. For years he had made no attempt at stealth in his comings and goings, since there had manifestly been no need to take the trouble.

Now, finally, someone—or something—had come. These were not the stragglers and misfits he had expected, traveling on foot or riding patched-together snowmobiles and armed at most with nothing more powerful than an automatic rifle. Whoever commanded this craft showed every sign of having technological resources beyond anything he had ever heard of, and the size of the thing suggested large numbers of people aboard. He was not going to chase these people off with the battered hunting rifle and meager supply of ammunition that he kept at home.

He knew that he had better hope they were friendly. If they were actively hostile he would want to start running fast and hiding well. He wished that the warm spell were over, so that fresh snow would come and cover his trail.

At the present, though, his trail was there, and there was nothing he could do about it. The newcomers might well be friendly or at least neutral, and he was not going to carry caution to the extreme of fleeing without further investigation; he was comfortable in his present home and did not care to be chased out of it needlessly. Therefore, the best course was to get a look at the ship and learn what he could. He began trying to devise a way he could watch it without making his presence known or too easily discovered.

He would want to get away from the trail of footsteps without making new ones just as obvious; that was easy enough, he decided after a moment's thought. He could just climb up on the roof of the building, as he had before; the ship's crew might well not think to look up there.

Where to go from there was not quite so simple to determine. The snow was deep enough to tunnel in, but that would be incredibly slow, and the tunnels would fall in fairly quickly if the temperature remained above freezing; therefore, he couldn't go under the snow. He had no way of going above it, other than walking through it leaving tracks. But, although he could scarcely go around all the snow—over the years it had gotten just about everywhere—he could go around the open area where his footprints would stand out. He could crawl along under the pine trees. The snow-laden branches would hide him, and, though he would be leaving a trail, it would not be easy to spot.

There were pine trees bordering three sides of the parking lot, and more around the edges of the golf course; he wasn't sure if these had been intended to discourage stray balls or to keep such plebian objects as the back of a supermarket out of sight of the golfers. It didn't matter much what their original purpose had been; small and storm-battered as most of them were, they would still provide satisfactory cover.

He glanced back up over the roof and saw nothing but snow; he paused and listened, but heard nothing save the sound of his own body settling into the snow. That was hardly decisive evidence of anything, but he guessed that he had a few moments at least before anyone came close enough to notice him.

That decided, he slid back down the drift and reentered the store. His cola stood on the checkout where he had left it; he bolted the rest of it. There were still a few candy bars beside the cash register; he snatched up a handful and stuffed them in his pockets. Thus provisioned, and unable to think of anything else he would need that wouldn't weigh him down unnecessarily, he stalked carefully out and marched back up the drift once again.

When he was on the roof he paused, debating whether he should head for the pines immediately or take the time for another look at the ship. Curiosity won out over caution; he crawled up the snowbank and peered over the top.

The ship sat there, the color of raw steel, harsh and gray against the pristine white of the snow. He could still see no sign of a hatch anywhere, but half a dozen figures had emerged somehow and were milling about in the open space around the ship's bow, where the fourth green had once been. They gave every appearance of being human; they were upright bipeds of not more than two meters in height, each with two arms and a head. He could not distinguish much detail over the intervening distance, but they looked very ordinary, wearing matching blue coats and dark pants, as best he could tell.

He felt a certain disappointment that there were no tentacled horrors or twelve-legged lizards anywhere to be seen. Though it might still be conceivable that the thing was a starship, he no longer believed it to be one.

He continued to stare, watching especially closely when the newcomers moved out of the huge vessel's shadow. They were certainly human, and in fact they all appeared to be white male adults of similar build.

Such homogeneity depressed him slightly; he had had idealistic hopes that the great upheavals of the southward exodus might have uprooted old patterns of behavior, and that the survivors might have abandoned their old prejudices. It appeared they had not. He wondered whether there was a single woman, black, or Oriental aboard the entire ship. There had been a flurry of neofascist propaganda during the final chaotic period before everything fell apart completely and the area was abandoned; the apparent popularity of the pronouncements on racial purity and culling of the unfit had been a major influence on his own decision to stay where he was. Had such theories been carried beyond the propaganda stage? Perhaps there were no blacks or Orientals left; the northern whites might have driven them out of their own lands.

That, he decided, was unreasonably pessimistic. The fact that the men he could see all happened to be white might just be coincidental. Or perhaps some sort of discrimination was practiced; there was no need to assume genocide.

Perhaps whoever had sent the ship was afraid that neofascists still lingered in the wastes, ready to shoot any nonwhite they might see.

Perhaps there had been a struggle and the blacks or Orientals had won, and these whites were being exiled. The similarity in their attire he had at first thought to be military or paramilitary in appearance, but it might as easily be prison uniform.

He saw no sign of guards, however, and the men were behaving more like soldiers than prisoners, moving casually and looking about themselves. He told himself that it would be foolish to make any assumptions about their status until he had more information. Guesswork would get him nowhere.

The newcomers were gathering together now, forming a tight little knot. They appeared to be talking; he saw gestures, but could hear only faint and garbled traces that did not come anywhere close to being intelligible.

Something appeared to have been settled; they turned toward him and started making their way along the fairway in his direction, floundering awkwardly through the deep snow. His first thought was that this was an ideal opportunity for a closer look at them; he need merely remain out of their sight. Then a vague memory stirred; weren't there such things as infrared trackers that could home in on a person's body heat even over so great a distance? Hadn't he once heard of sound detectors that could pick up a human heartbeat a kilometer away? It could, of course, be coincidence that the group was heading in his direction—but he usually believed in the old adage, "Better safe than sorry." It was time to leave the roof. If the advancing party of men changed direction to follow him, it was time to leave the area as quickly as he could.

He had no proof that they were after him, or that they thought anyone might still be alive out here; they might merely be interested in the old market. Or perhaps they had seen his trail from the air and wanted to investigate, not knowing whether he was still in the vicinity or not. They might be honestly friendly; he saw no rifles or other recognizable weapons in evidence. Three of the six did carry various pieces of equipment, none of which he could identify, and other gadgets were hung on belts. One man held a boxy contraption out before him and glanced at it every so often.

That box might be a tracking device or something else entirely; Starkman had no way of knowing.

Cautiously, he pushed himself back down toward the edge of the roof, behind the top of the drift he had been peering over. Once he was reasonably sure he was out of sight, he shifted into a crouch and hurried to the corner of the building. From there he inched his way along the side, watching carefully. He was more than halfway to the rear edge of the roof before he could see the advancing strangers; the heaped snow screened him perfectly.

If he were to avoid leaving a trail of footprints he had to go from the roof directly to the cover of the pines; the only trees directly adjoining the building were along the back, the side nearest the strangers. He had to risk being spotted while crossing the intervening distance.

When he glimpsed them over the snow he dropped flat and began crawling, moving as quickly as he could. By the time he reached the rear edge he could see the newcomers even while prone; he hoped they would not notice him.

There was a fairly tall pine close to the corner he was approaching; in a quick dash he ducked behind it. He waited, listening, but heard nothing except a desultory mutter of unintelligible conversation and the shuffling of boots through wet snow. Carefully, still listening, he turned and lowered himself down over the edge of the roof. After a moment's hesitation he let himself drop.

There was a branch in his way, hidden by the snow; it snapped loudly as it broke beneath his weight. He landed on his back atop the branch and a pile of snow, his breath jarred out of him. More snow pattered onto his chest from the remaining tree limbs. He lay utterly still even when his breathing managed to restore itself.

The faint crunching of footsteps ceased and the voices became quicker, louder and more excited, but still incomprehensible. He guessed from the sound that the party was about fifty meters away.

Much sooner than he had expected someone called loudly, "Hello! Where are you?" These people, whoever they were, reached decisions quickly; they had taken only a very few seconds to debate before calling out. They might be following a standard course of action, or one might be sufficiently authoritative or reckless to take the matter into his own hands, but Starkman was quite sure that had roles been reversed he would not have called out like that, with no idea of whom they faced, nor even of how many.

But then, he realized, if their instrumentation were good enough, they might already know that they faced a lone unarmed man.

In any case, they were apparently trying to seem friendly. Still, he had no intention of simply walking out into the open. It could be that they were fearless due to absolute confidence in their superiority, rather than because they were actually friendly. He lay still.

"Listen, we mean you no harm! We're here to take you back south with us!"

The strangers spoke English, and with an American accent—or at any rate the two he had heard so far did. They, or their parents, were presumably his former countrymen.

There was still plenty of room for speculation. Were they being exiled after defeat at the hands of native southerners, or had they taken over somewhere and set up a new United States? Were they telling the truth? If so, why did they want to take him south? Was it simple humanitarianism, or were they planning to reoccupy the country and seeking whatever information he might provide about current conditions?

He wondered if he were being unreasonably suspicious. Perhaps he was becoming a trifle paranoid after living so long alone. After all, these men were obviously civilized, and were more or less his own kind.

There was the catch; he was more or less, but not exactly, like them. Many of the reasons he had had for not going south in the first rush, fifteen years ago, or in any of the later parties, still applied. He had no way of knowing what conditions were in the southern lands, who was in charge, how things were run. The fact that the entire group was white and male was not an encouraging sign. Even if the neofascists weren't in power, even if the ruling elite, whatever it might be, was willing to tolerate him, he did not want to go through life merely being tolerated. If there was a truly free and pluralistic society, one that would accept him, he might be better off going, but until he had some assurance that that was what awaited, he preferred to stay where he was. He fished his sunglasses out of his pocket and put them on, being careful not to disturb the snow and risk another fall, and trying not to rustle his coat too much. With the glasses securely in place he twisted around, still attempting as much stealth as he could, and looked out toward the strangers.

They were all looking almost exactly in his direction; the one with the box was holding it out toward him, it seemed. They were perhaps fifty meters away.

Starkman rolled over and began crawling away from the supermarket, staying out of sight beneath the pines. Five of the men kept on as they had, but the one with the box suddenly stopped and called, "Hey, don't go away!"

They were tracking him. That had become obvious. He stopped, wishing he had a weapon, and tried to think what he could do.

"Hello! Are you there?" one of them called.

He slid down behind a thick clump of branches and the concealing pile of snow it supported.

"Come on out!" a new voice shouted.

"Do you see him anywhere?" They were close enough for him to hear their conversation as well as their cries.

"What if it's a girl?"

"It doesn't matter, does it?"

"No, but you said 'him.'"

"So what? You think you're smart, don't you?"

Starkman listened, and was suddenly confused. Was his memory playing tricks on him? These people sounded more like a bunch of kids playing hide-and-seek than an adult search party; was that accurate, or was he misremembering the way people spoke? He had heard no voice but his own in so very long that he could not be completely certain. Perhaps he was confusing ordinary conversation with the stilted dialogue in his books.

The man with the box pointed at him and announced, "He's right there."

"I see him! I see him!" another one shouted.

For a brief moment, Starkman considered flight. He was more accustomed to snow than these men were, that was obvious. However, it was also obvious that they could track him; even if he were to escape temporarily they could hunt him down at their leisure. Furthermore, his chances of escaping even temporarily were not really very good; he was entangled in the pines, for one thing. If he emerged from the trees on their side he didn't have a chance, one against six; if he moved along the row of trees they could easily pace him in the clear and catch him when he emerged. If he were to make a break back out across the parking lot he might have a chance, as they would be delayed by the pines—but it wasn't a very good chance, and if they were armed they might just decide to shoot him in the back if he ran.

Reluctantly, he raised his hands above his head and stood up, pushing his way out through the snowy branches. He looked over his captors from behind his mirrored glasses.

They were not aliens. Quite aside from their human features, they wore perfectly ordinary fleece-lined coats, blue, with heavy steel zippers and drawstring hoods. The hoods hid their hair, and Starkman wondered idly how long they wore it; his own hadn't been cut in years, and was washed only rarely since heating water was such a nuisance, so that it reached halfway down his back in greasy, uncombed strands. The blue coats bore no insignia, and were not overtly military in style, but they did all match—as did the blue leggings and black boots each wore.

All six appeared to be young men, in their early to middle twenties, and all wore full beards, rough and untrimmed. That seemed slightly odd; even if beards were the current fashion, he would have expected at least one of the six to take the trouble to maintain a neat appearance by keeping his trimmed. It was curious that all six were slobs.

The colors of their beards varied, which comforted him somewhat; during the chaotic period at least one radical faction had advocated the elimination of people with black hair or other signs of impure heritage. Two of the six he now faced had thick black beards—but all had pale white skin.

All were in fact pale to the point of being pasty and unhealthy in appearance; Starkman wondered how they managed it. Were even the southern lands so cold and sunless that no one was tanned?

There was something slightly peculiar about their eyes, as well; Starkman wasn't sure what it was, but they were somehow not quite right. After a second's thought he dismissed it as either his imagination, or faulty memory from having lived so long without seeing any eyes but his own, or a subtle sign that these people were drugged. The last possibility, he realized, might account for the slightly bizarre behavior they had displayed.

"Hi," one of the men said. "We're here to take you to Brazil."

"You can put your hands down," another added helpfully. "We're not armed."

"You're not supposed to tell him that!" snapped the one who had spoken first.

These people, Starkman told himself, were acting like half-wits; it was not just faulty memory on his part. Either they were really as stupid and childish as they appeared, in which case it seemed incredible that they could be in command of a craft such as the one that had brought them, or they were playing dumb to lure him into overconfidence.

"You're from Brazil?" he asked.

"Naw," replied the one who had suggested he lower his hands. "That's just where we set up our headquarters here on Earth. Actually, we're from the Galactic Empire."

Chapter Two

Starkman said nothing for a long moment as he considered the stranger's claim. Cautiously, he allowed his hands to drop to his sides.

These people were ordinary human beings. It was utterly impossible for him to accept them as extraterrestrials. Therefore, either he was finally going insane from his long isolation—perhaps this entire episode was an elaborate hallucination—or he was dealing with a bunch of lunatics.

Or, just possibly, these men were perpetrating some complex hoax; he could imagine no reason for them to do so, but he had to admit it was a possibility.

"You're from the Galactic Empire?" he asked at last.

"Yeah." It was the one who had first mentioned Brazil who replied; he seemed to be more or less the group's spokesman.

"Sort of," added one with a blond beard.

"Sort of?"

"It's kind of complicated; don't worry about it. Is there anyone else around here? We're supposed to take everyone we can find back to Brazil with us; that's why we're here."

Hesitantly, Starkman asked, "Who are you guys?"

"I'm Joey," answered the black-bearded spokesman. "That's Bobby, and Jim, and Jason, and Bill, and Mike. We're the search party."

"The search party from the Galactic Empire."

"Yeah, well, one of them, anyway. Who're you?"

"My name's John Starkman." There could be no harm in telling the truth.

"Pleased to meet you, Johnny." Joey stuck out a hand; reluctantly, Starkman shook it. "Is there anyone else around here? We're supposed to get everybody we can find. We saw the tracks in the snow over there, so we landed, but we didn't see anyone but you with the scopes."

After a second's hesitation, he admitted, "I'm alone; there hasn't been anyone else around here for years."

"Maybe he's lying," said Jim, the blond who had spoken before.

"Why would he do that?" asked Bobby.

"To confuse us; maybe he thinks we're bad guys, and he's trying to protect his friends."

"Aw, that's silly. You're just a worry-wart," replied Jason, the one who had told him to lower his hands.

"I didn't see anyone else with the scope," Joey said.

"Try again," Bobby suggested.

Joey shrugged, and fiddled with the box he held. "There's just us," he reported.

"Okay," Jason said, "Then he was telling the truth. Come on, let's get back to the ship; I'm getting cold."

"Sissy," Jim muttered.

"Come on, Johnny," Joey said, gesturing to Starkman.

"Where to?"

"To the ship, of course, so we can take you back to Brazil." He evinced surprise, but no trace at all of annoyance or hostility.

Peculiar as the search party's behavior might be, there was no sign of coercion, no implication of force; Starkman felt sure enough of himself to ask, "Why?"

Totally befuddled, Joey replied, "Why what?"

"Why do you want to take me to Brazil?"

"Because that's what we're here for!"

"But why should I come with you? Why would I want to go to Brazil?"

"Huh?" Joey stared at him. "I don't know; 'cause it's warm, I guess, and there's other people."

"What if I don't want to go?"

All six were staring in open-mouthed amazement. "You've gotta go!" Joey yelled. "You're supposed to *want* to! Don't you want to see Brazil?"

"Not particularly."

"Look, you gotta come with us! If you don't we'll get in trouble, and I bet you'll get in trouble, too."

"Get in trouble with whom? Who sent you here?"

"God, you're suspicious! The Galactic Empire sent us, of course, and we could get in trouble with the Governor. Who else?"

"I never heard of the Galactic Empire, and I don't know what governor you're talking about."

"Oh." Joey was momentarily silenced.

"Well, yeah," he said after some thought, "I guess you wouldn't, living out here in the snow like this. I guess you haven't even got a TV, or anything, do you?"

"Not one that works." Starkman's throat was beginning to feel dry and tired; he hadn't spoken so much in several years. Joey's voice seemed unnecessarily and painfully loud, as well.

"Well, see, the Galactic Empire runs everything. They run the whole world here on Earth, and all the rest of the galaxy, too."

"But who tells you what to do?"

"Us? Oh, we work for the Governor!"

"Who's the Governor?"

"He's the boss of Earth, of course! He runs everything here. It was the Governor who brought us all to Earth in the first place."

"What's this governor's name?" Starkman's continued questions served two purposes; he was gradually compiling Joey's story, in order to have some idea what he was up against and to be able to spot contradictions, but he was also delaying in hope that something would arise to prevent his abduction—despite their outward innocence, he had little doubt that he faced abduction.

"I don't think he's got a name; he's just the Governor."

"Everybody's got a name," Starkman insisted.

Joey looked puzzled. Before he could come up with a response, Mike spoke for the first time.

"I'm gettin' cold, Joey," he said. He had an unpleasant tenor whine.

"So'm I," Jason agreed.

"I'm goin' back to the ship," Mike said. He turned and began trudging off through the snow.

"Hey, come back here!" Joey yelled.

"If he don't want to come, I'm not gonna stand around and argue," Mike called back. "He can stay here and freeze if he likes!"

"Oh, hell." Joey looked about and saw that his four remaining companions were obviously on the verge of following the deserting Mike. "Mister Starkman, you *gotta* come with us!" he pleaded.

"Nope," Starkman replied, "I'm staying here."

"Maybe we should drag him," Bobby suggested.

There was halfhearted agreement from the others.

"Okay, you guys grab him," Joey said.

Starkman braced himself for a struggle, but no one moved.

"Come on!" Joey insisted. No one moved, himself included.

"You do it," Bobby said at last. "He's bigger than me."

This was not at all obvious. Starkman would have guessed that Bobby was a pretty even match for him, perhaps as much as two kilos heavier than himself, with similar height and build but better fed. He suppressed a grin; he might, he thought, have a chance after all.

"Yeah," Joey said, "but there's five of us!"

"You do it," Bobby repeated.

Joey looked around and saw no sign of cooperation; the others averted their eyes rather than meet his gaze. "Okay," he said at last, "but you guys have to help me drag him."

There was a lukewarm chorus of agreement. Joey stepped forward and reached out for Starkman's arm.

Starkman pulled the arm away and brought his other hand up in a fist. Without really meaning to, he caught Joey with a good solid blow under the chin, knocking the young man backward; he landed hard, sitting in the snow.

Starkman stepped back and raised both fists, ready to strike out when the others jumped him, but the attack he expected never came; instead the other four backed off and watched.

Joey sat rubbing his throat; he looked up at Starkman, and his expression shifted from astonishment to rage. "You hit me!" he said.

Starkman said nothing.

"I'm gonna kill you!" Joey screamed; he surged up onto his feet and charged forward, both fists swinging wildly at air.

Starkman laid him out flat on his back with a simple punch straight to the nose, and felt a twinge of pity and guilt; the kid had left himself wide open, putting up no defense at all. Joey's own blows, the very few that landed, had no power behind them; Starkman had been hit harder by the snow that fell on him from the pines.

Joey sat up again, a trace of blood running from his nose and vanishing into his bristling mustache. "I'll get you!" he whined. "I'm gonna go and tell the Governor and come back here and kill you! You shouldn't've hit me like that!"

Starkman remained silent. He stood, fists clenched, almost trembling with an unspent rush of adrenaline, as Joey got to his feet and limped off toward the ship. The remaining four looked from Joey to Starkman and back; it was plain that they would soon follow their comrades and leave him alone. Starkman could see that Mike had almost reached their immense vessel. He intended to stand where he was and watch them leave, like any animal defending his territory; once they were all back on the ship he could relax.

Before any of the four could take a step, however, a voice spoke from somewhere, saying calmly and clearly, "Wait, please."

Startled, Starkman looked about wildly. There was no sign of anyone but himself and the search party.

The voice spoke again, still calm; it had an air of careful control, and was pitched so that he could not decide if it was male or female.

"Mister Starkman, please reconsider."

"Who's speaking?" he demanded.

"This is the ship's commander."

"Where are you?"

"I am on the ship."

Starkman relaxed slightly, and chided himself for

reacting so strongly to what was apparently just a so-phisticated public address system. The speaker could hear him, so there was some sort of directional microphone involved as well, but it was still nothing supernatural.

"What do you want of me?"

"We want you to come aboard the ship and accompany us to our headquarters in Brazil. We mean you no harm."

The voice was inhumanly placid. Starkman wondered what kind of person the commander was—if it was actually the ship's commander speaking. He had a moment of wild fancy that the commander wasn't human at all.

"Why?"

"We are carrying out our instructions. I do not know the reason we were so instructed. I do know that no harm will come to you if you cooperate. Our superiors do not wish to harm any human being. However, we have been instructed quite emphatically that we are to bring back every person we find in this area, with no exceptions. If we allow you to remain here we will not have done as we have been instructed."

Starkman was sure that the speaker's native language was not English; either that, or he or she had been reading from a prepared script. No one accustomed to speaking English would have phrased his speech so precisely and delivered it in so steady a way.

"I don't care whether you do as instructed or not," he retorted.

There was silence; the voice did not answer him.

When the silence continued and became awkward, the four remaining young men began shuffling their feet, snuffling and glancing at one another, but none of them made a move either toward Starkman or toward their ship.

Finally Starkman's patience ran out. He turned away and started to leave.

Before he had completed his first step there was a fierce hissing, and a cloud of steam rose around him. He froze, then looked at the ground whence the steam had come.

Two neat grooves were cut in the snow, one on each
side of his feet. Each was a centimeter or so wide, and
deeper than he could see readily. Tracing the lines forward they ran on into the row of pines; he could see a
wisp of smoke curling up where one intercepted a low-
hanging branch. Following them back they ran straight
through the group of strangers, dividing them two to
each side, and pointing directly back toward the ship.

The calm voice spoke again. "We cannot allow you
to leave, Mr. Starkman. That was a warning shot. If
you attempt to depart again I will be forced to use the
same weapon on your feet. I have been forbidden to kill
any human being, but I am no longer forbidden to injure
or incapacitate if necessary. Please do not continue to
be uncooperative."

Starkman looked at the slices through the snow. He
had seen no light, but he guessed that they had been
made by a high-powered laser or something very similar. Even if his feet were more durable than snow, he
suspected that they could be very effectively flash-fried.

"What about my home? Can I bring my belongings?"
If he were allowed to move beyond the pines, on any
excuse, he thought he might be able to make a break
for it and get beyond the weapon's range.

"I am sorry, Mr. Starkman, but we cannot take the
time to fetch your belongings. I am sure that no one
will disturb them, and I have no reason to believe that
you will not be permitted to return here in the future.
Should you prefer not to return in person, perhaps
someone could be sent to fetch them. At present, however, we cannot do so."

Frustrated, he turned back toward the ship. He was
not accustomed to being thwarted; for several years now
he had lived alone and always had his own way. It was
aggravating to realize that he could no longer rely on
ever again having his own way.

"Damn," he muttered. He began stalking through
the snow toward the ship, taking what amusement he
could from stamping in the grooves the weapon had cut
in the snow for as far as they ran. Twenty meters from
where he had stood they rose up out of the snow and

vanished, and he was left without even that inadequate outlet for his emotions.

The four young men followed him, floundering in the snow.

Joey and Mike had already gone aboard, and the other four were unable to keep up with him, so Starkman found his way by following their tracks rather than being guided. He did not pause when he first reached the side of the ship; he was in no mood to admire the immense machine that towered over him. Instead he marched on, glowering down at the snow. He followed the tracks for the entire length of the craft—it was at least two hundred meters, he judged—and on to the base of a ramp that hung down from the underside of the ship's bow.

At the foot of the ramp he paused, and it sank in for the first time that the ramp itself was immense, in proportion with the ship. It was a dozen meters across and twenty meters long. Since the main body of the ship was lying flat on the ground, not raised up, only the curvature of its hull raised the ramp's upper end above the lower. This provided an extremely shallow slope. There were no joints, seams, welds, or rivets visible anywhere. The upper surface was flat, a single sheet of grooved metal, and the underside was curved so that when the ramp was raised it would blend evenly into the hull; as a result its thickness varied from half a meter to more than a meter, yet it all appeared to be a single solid chunk of metal.

Starkman looked at the ramp and found himself remembering old pictures of tanks being unloaded from landing craft, and jeeps driving out of the noses of aircraft. The hatch at the top of the ramp was in scale with the rest, and more than wide enough for such vehicles. A small army could march up and down such a ramp and through such a hatchway without breaking formation.

He recalled his earlier guess that the vessel might be a warship of some kind. What but a military transport, he asked himself, would need so massive a ramp and so broad a hatch? The laser weapons that had boiled the snow around him were proof that the ship was armed.

What, he wondered, was he getting into?

The four others came up behind him. One of them—he had already forgotten their names, but he thought it might be Jason—said, "Well, what are you waiting for? Get on board!"

Reluctantly, Starkman walked up the ramp, looking about himself.

The well into which the ramp would lift was all smooth, bare metal; there were none of the labels or machines he had half expected, nothing to give a human scale anywhere, nor any sign of the builder's nature or native tongue. He felt dwarfed by the blank, shining expanses. The ship could have been built for elephants as easily as for humans, from what he could see of it.

The hatchway at the top slid open as he approached, revealing Mike and Joey waiting in the corridor beyond. As Starkman walked deliberately close to him, Joey shied away as if expecting another punch in the nose; Starkman resisted the temptation to shout "Boo!" in his face and watch him jump. Mike noticed the action and gave his companion a look of disgust.

"Wipe your nose," he said; then he turned to Starkman and gestured. "This way," he commanded.

Starkman wondered how Mike had so quickly become Joey's superior; he watched as Joey awkwardly brushed at a slow oozing of blood that was clotting in his mustache.

With Mike in the lead and Joey bringing up the rear, he was escorted down the vast empty corridor. It was six or seven meters high and almost as wide; Starkman wondered why so much space had been wasted on a simple passageway aboard an airship—if that was what this was—where weight and space were presumably at a premium.

The floor and walls were all bare metal, completely unadorned; light was provided by three strips of glowing white plastic, one along the center of the ceiling and one along the top of each wall. Starkman was unable to determine the nature of the light, whether the plastic strips were merely covers or whether they themselves glowed; they were all far overhead, where he could not get a good look at them.

There were hatchways along each side of the corridor at infrequent intervals, and one formed the end of the passage; each was, like the corridor, six or seven meters in height and five or six meters in width. Starkman had a curious feeling of having shrunk; he imagined himself a mouse in a warehouse as he passed by the first of the gargantuan portals. If he had not already convinced himself that his captors were as human as he was, he might have believed that this ship was designed for aliens—very large aliens.

As they approached the second hatchway on the left it slid aside, though he had seen neither of his two escorts do anything that might have triggered it. Mike stopped before the open door and gestured for Starkman to enter.

He stopped and gazed in.

The room beyond the hatchway was large, easily a dozen meters square, and its high ceiling, fully as high as the corridor's, gave an impression of vastness. The walls were more of the bare, plain metal, and strips of plastic provided light on all four sides; ceiling and floor were lined with a thin layer of something slightly spongy. There was a door in the center of each of the four walls; though only the one through which he peered was so absurdly large, the others were not quite normal, each being about four meters square. The remaining wall space was lined with ordinary steel-frame, black-painted cots, and two neat rows stood in the middle of the chamber as well, for a total of thirty or forty beds. About a third showed signs of recent use; the rest were all neatly made up with taut white sheets and drab gray blankets. The room was as colorless as the landscape outside.

It was occupied, however, and by people nowhere near so drab.

There were perhaps a dozen of them, clustered on three or four of the beds in the center; Starkman had just begun to look them over when Mike, in the proper Hollywood-movie storm-trooper fashion, ordered, "Get inside."

Starkman stepped into the room with only a brief glance in Mike's direction; the door slid shut behind

him, leaving Mike and Joey in the corridor outside and
himself in the room with a group of complete strangers.

They looked nothing like the uniformed and healthy
young men of the search party; for one thing, none of
them were in their twenties. Starkman counted five
children, ranging from an infant at its mother's breast
to a girl of about thirteen; there were eight adults besides himself, five men and three women, all at least
thirty-five, he guessed, and one old man who appeared
to be well into his eighth decade. That made Starkman
the youngest adult present by a year or two, he judged.

The gathered people wore a motley assortment of
clothing, much of it as battered and worn as his own
decrepit coat, and hair styles tended toward the di-
sheveled and uncut. Starkman was pleased to see that
one of the men was a black in his forties; it appeared
he need not have worried about genocide—at least, not
so far as blacks were concerned.

There was a moment of awkward silence, broken at
last by the black man. He rose from where he sat and
said, "Hello; my name's Jerry White."

"I'm John Starkman." He took a cautious step forward.

White met him halfway, and they shook hands.

"I guess the zombies picked you up the same way
they did the rest of us?"

"Zombies?" The term obviously referred to the uniformed young men. "I guess so. Why do you call them
that? They don't look dead to me."

"It's not my idea, it's Jenny's." He gestured toward
one of the women. "They're about as stupid as zombies,
and they all look alike—and we haven't got a better
name for them."

"Oh." Starkman had no better name for them, either,
and the label "zombies" was more convenient than his
own of "strange childish young men."

"Let me introduce you around; it'll give me a chance
to make sure I've got everybody right myself," White
suggested.

Starkman agreed readily.

The nursing mother's name was Althea Vandeventer; the baby, Lazarus, was her youngest at three months

old. Ruth was her oldest, at thirteen, and Joshua second, at six; there had been other children who hadn't survived, her husband Peter Vandeventer mentioned. The family had been living on his ancestral farm near Washington, Pennsylvania, since they had married, before the cold first became serious, and they had refused to give up their land, though of course nothing had grown there in years except in their improvised greenhouse. After a certain point it had been little more than habit that kept them there, and with the new child's birth and Ruth reaching puberty the arrival of the great ship that had landed yesterday in what was once their cornfield had seemed like a gift from God, sent to fetch them away to somewhere their children could live a normal life.

The other two children were Charlie and Kathy Saslov, aged nine and eleven respectively; they and their mother, Jenny, had been working their way south from Englehart, Ontario, since the death of their father more than a year earlier. They had been spotted and picked up two days ago outside Pittsburgh. Jenny had been reluctant to cooperate at first, but knew she could never escape burdened with two children—and she wasn't about to abandon them. Kathy was scared; Charlie thought it was all "neat."

Janet and Wesley Hatfield were from West Virginia, in their forties, and still too confused to have much to say. They had been roused out of their beds the preceding night by the sound of the ship's engines. Janet claimed she had thought she was dreaming, and had gone along aboard the ship because what harm could it do, in a dream? Wesley wouldn't say even that much.

The old man's name was Robert Carvel, and he had had all of downtown Pittsburgh to himself until two days earlier, when the ship had set down in Point Park and the six zombies had hunted him down in his hideyhole in the old William Penn Hotel. He wasn't at all happy about it, but he hadn't dared argue; his old bones were brittle, he said.

The last of the group to speak was Nathan Molley, sixty-one years old, who had stayed on in Good Intent, Pennsylvania, out of sheer stubbornness. As he explained it, no one had ever asked him to leave until yesterday evening, when the polite young men came and invited him aboard. Once someone *asked* he was glad to go.

Starkman asked Jerry White about his own origins; he had, he said, been the first to be picked up on what was apparently an east-to-west route, since he had been living in the Monroeville Mall, east of Pittsburgh. He'd been getting low on supplies, and was tired of moving on from one wasteland ruin to another, so after some hesitation he'd boarded the ship willingly.

Thinking it over, Starkman realized that the others had, indeed, been picked up more or less in a line moving southwest—assuming Good Intent, which he'd never heard of, fit—but the ship must have doubled back, since he had been living in Pennsylvania, not West Virginia. He had moved around over the years, and was unsure just where he had been picked up, but his home had always been in the suburbs south of Pittsburgh. He had had no idea there were so many people living in the area; he had twice made the journey into the city, but had never found any evidence of continued human habitation there. He had preferred the outlying areas. He had had an irrational fear that the tall buildings were about to collapse on him, and the echoes in the empty streets had made him nervous.

When everyone had been introduced the conversation began to sag badly. In an attempt to shore it up, White remarked, "Looked like one of the zombies must have taken a fall and bumped his nose."

Starkman could not resist boasting. "It wasn't a fall that bloodied his nose," he said, holding up his fist.

"You did that?" Carvel asked, evidently delighted. "Good, good! Serves him right!"

"Why did you hit him?" Peter Vandeventer asked.

"I didn't want to come; I was happy where I was."

"They forced you to come?" White asked.

"More or less; they threatened to drag me, so when one of them tried it I hit him. The poor fool couldn't fight any better than little Lazarus here, but then someone in the ship fired a warning shot about a centimeter from my feet, and I decided there wasn't any point in getting fried."

"What kind of a warning shot?" Jenny Saslov asked.

"Laser, I think; something that boiled the snow, anyway."

"I guess it's just as well I didn't resist, then—but I'm glad that someone did!"

"It was a damn stupid thing to do," Molley said. "One look at this ship should have been warning enough for anyone. Besides, they don't mean us any harm."

"If they don't mean any harm, then why did they bring Mr. Starkman here by force?" Jenny Saslov demanded. "Why did they insist I come with them? Why did they pick up any of us?"

"To take us south, of course!" Molley replied.

"But why?"

The conversation degenerated rapidly into an argument. Starkman was not interested; no one here knew anything more than he did, apparently. He seated himself on a cot and looked around the room. White sat down beside him.

"Why don't you take off that coat? You must be hot in here."

It was true that the room was warm. Starkman looked down at himself, then shrugged. "I'm used to it," he answered, but he reached up and pulled down the zipper.

"Suit yourself. I don't think you have to worry about anyone stealing it; none of us is going anywhere. The corridor door only opens to let in new arrivals. If you change your mind, just stick it under whichever cot you pick." He pointed out where the other captives had put hats, coats, scarves, boots, a few packs, and even a rifle.

Starkman peered at the gun, and White guessed his thoughts.

"It's Jenny Saslov's, and there aren't any bullets; she kept it to scare people off. It didn't scare the zombies at all."

"Oh." The excitement of his capture was wearing off, and he was becoming aware of a need. After a moment's hesitation, he asked, "Is there a bathroom around here?"

"Oh, yeah, of course! That door there." He pointed

to one of the broad hatches. "Just tell it to open; it's voice-actuated. There's a sort of a shower, too, if you want one."

Starkman wasn't sure if that was meant as a hint or not, but he didn't much care; he knew he was rather malodorous, even if nobody had said anything about it yet, and the thought of a real shower was extremely attractive. "Thanks," he said. He stood and crossed to the indicated door. After a moment's hesitation, he said, "Open up."

The door slid aside, revealing another bare metal chamber, this one considerably smaller. It appeared to be a featureless cube, about four meters on a side; it was lit by glowing spheres set in each of the eight corners. Starkman saw nothing resembling plumbing anywhere, nor any other doors, but, looking closely, he could see that there were several tiny holes in the nearest wall. Cautiously, he stepped inside; the door slid closed behind him.

The light from the eight spheres seemed very peculiar, coming as it did from below as well as above; he wondered if there were any way to shut off the lower four. He also wondered what kind of lunatic had designed this ship.

There was a drain in the center of the floor, about half a meter across. Looking about, he saw that all four walls and the ceiling were patterned with tiny holes, and that there were some even in the floor. There were, however, no other signs of plumbing.

"Damn," he muttered. He should have asked White how the place worked.

"May I help you?" asked the voice that had identified itself earlier as the ship's commander.

Starkman paused for a moment, then said, "What are you doing? Are you watching me?"

"No. The privacy conventions of your society have been explained to me. However, all mechanisms aboard this ship, of whatever nature, are under my control; therefore, I continue to monitor the audio circuits throughout the ship, in order to be able to provide whatever services may be required."

"Every mechanism is under your control?"

"Yes."

It was immediately obvious that nobody, no matter how bizarre they might be, would have designed a ship this size where everything down to the plumbing was controlled by a human commanding officer. "Are you a computer?" Starkman asked. "I thought you were human."

"I am not human. I am not exactly a computer, however. I am a conscious entity artificially created for the purpose of running this ship."

"You're a machine, though?"

"That is subject to debate. Your language is not sufficiently precise in its definitions for a definite answer."

After a moment of groping for his next question, Starkman decided to drop the subject temporarily, and inquired instead after the workings of the plumbing.

The drain in the center of the floor rose up on a shaft of gleaming metal and irised open.

This was not like any facility Starkman was familiar with, but it served its purpose. When he had done with it, he asked how the shower worked.

The voice directed him to a panel in one wall that slid aside as he approached, where he could store his clothing while he bathed; he stripped and put his worn, dirty garments in the opening. The panel slid shut, and a fine spray of water jetted from the holes in the ceiling—and also from those in the walls and the floor.

The water was gently warm, but having it strike him from all sides at once was a novel and not particularly comfortable sensation at first; he blinked, and instinctively backed toward one wall, avoiding as much of the spray as he could.

"Temperature and pressure may be adjusted, and soap added, at your request," the calm voice informed him.

"Can you shut off the water from the floor?"

The upward spray ceased.

The horizontal jets of water he could deal with, and even enjoy after he'd had a moment to adjust to them; he tinkered with the water temperature, and the nature of the spray, wondering idly how the controlling entity—whatever it was—could hear him over the hissing and splashing of the water. He tried the soap briefly, but didn't like the feel of it. At last he obtained a fierce fine hot spray, and luxuriated in the resulting fog of steam and vapor and the sharp, invigorating feel of it on his skin.

After a few moments he felt cleaner than he had in months or even years. He ordered the water off and asked about a towel. Instead of supplying him with one, the spray of water was replaced with jets of warm air.

When he was thoroughly dry the panel in the wall slid aside, revealing his clothing. He hesitated to put the dirty garments onto his wonderfully clean body, but decided he didn't trust the ship to clean them without destroying them. He dressed, slung his coat over his shoulder, and told the controlling entity to open the door.

It obeyed, and he stepped out into the larger room. His fellow prisoners were somewhat more spread out than they had been, scattered across a dozen beds instead of clustered on a third that number. Jerry White sat nearest him; he rose, smiling, and began, "I guess you're..."

His smile froze, then vanished, as he took in Starkman's face. "Oh, my God," he said. "What's wrong with your eyes?"

Chapter Three

Starkman grimaced with annoyance as he realized that he had forgotten to put his sunglasses back on. When he had undressed he had stuck them in his coat pocket, as he always did, and when he had dressed again he hadn't put the coat back on, so that it had been easy to forget about them.

"There's nothing wrong with my eyes." He swung his coat around in front of himself and reached for the pocket.

"But they're yellow!"

"I *know* they're yellow. They've always been yellow, and they probably always will be yellow." He found the glasses and put them on.

"But why? What's wrong with them?"

"*Nothing* is wrong with them, dammit. I was born with yellow eyes, that's all, just the same as you were born with dark skin." He realized the others were all watching and listening, staring at him.

"My skin is perfectly normal. I never saw anybody with yellow eyes before."

"All right, so I'm a freak! Just shut up about it and leave me alone!" Starkman shoved him aside and stalked past. He picked himself a bed and sat down heavily upon it.

He mentally cursed his carelessness; he was an outcast all over again, it appeared. Always, he remembered, people had stared at him, and he had had to struggle to be accepted as a human being. Sunglasses had been all that made his life bearable in his self-conscious adolescence; he had hated cloudy days and

41

prayed for sunlight, making his mirrored lenses almost a fetish, keeping an extra pair always close at hand. Each fall he had had to convince a new set of teachers that an exception should be made for him and sunglasses allowed in the classroom; each time he had won his argument simply by taking them off once.

His few serious dates had all ended in failure; girls who had not known of his peculiarity in advance were invariably shocked and revolted when he removed his glasses, and those who *had* known of it but still dated him were always motivated by either pity, which he could not stand since there was nothing *wrong* with him, or by a morbid fascination that repulsed him as strongly as his appearance repulsed others.

There was nothing at all wrong with him, and the stories he had told of eyes too weak to take strong light had been pure lies. He simply had a unique pigmentation. Doctors had examined and analyzed it, and told him that it was a minor genetic peculiarity, harmless, and nothing to be concerned about. The doctors did not have to live out their lives with eyes that were rich golden yellow throughout both iris and ball.

It was his eyes, more than anything else, that had prompted him to stay behind in Pennsylvania when the snows came and everyone else fled south. The climate change gave him a chance to live alone, away from misguided pity, sick curiosity, or simple disgust, without having to hack himself a place in the wilderness. Furthermore, in the eugenics campaigns that were announced and pursued with varying degrees of enthusiasm during the exodus, he was quite sure that his little quirk would have been grounds for sterilization or death. He was not sure that he would have any real objection to sterilization—after all, he wouldn't want to stick anyone else with a life like his own—but he had no desire to risk death.

He sat on the bed and looked around at the other prisoners, and saw the old familiar expressions of fear, curiosity, and pity in their faces. It hurt; he had forgotten how much it hurt to be stared at.

He wanted to turn away, but stopped himself; if he turned away, he knew, he would be accepting the division between himself and the other, normal people. He returned their stares from behind his mirrored glasses.

The silence was broken by Carvel's voice.

"Aw, hell," he said, "so the guy's got funny eyes. That's no excuse to gawk at him. We're none of us quite normal, after all, or we wouldn't have been living up here in the snow."

"Speak for yourself," Molley retorted. "I'm not a freak."

There had been a time when Starkman would have taken umbrage at being called a freak, and done his excellent best to pound Molley's face in. He had gotten over that; he had finally managed to accept that he *was* a freak, literally. Still, the term stirred up old resentment and anger, and his hands clenched into fists as he fought down a rude reply. There was no point in fighting; it would only antagonize the others unnecessarily. He would win no friends by beating a sixty-one-year-old man.

Jenny Saslov answered for him, telling Molley to shut up. Her face was one that showed no sign of fear, but only pity and a tinge of curiosity. Molley's was a shifting display of anger, fear, and distrust. Carvel's was unreadable. White's expression was still nothing but surprise.

All four of the older children showed fear and curiosity mingled, but it was Kathy Saslov who came over and asked, "Can I see?"

Reluctantly, he removed his glasses and met her gaze. She blinked, turned away for a moment, then turned back.

"They're kind of pretty," she said.

The other children came up behind her, and he looked at them, one by one. Joshua ran away, but the others met his gaze and looked back.

"Don't feel too upset about Joshua," the boy's father called. "He did the same thing the first time he saw Mr. White."

"He'd never seen a black man before," Althea Vandeventer agreed.

"And none of us ever saw a man with yellow eyes before," Ruth added.

Starkman was still too full of hurt and anger to reply or even smile, but he managed a nod as he slid his glasses back on.

After a moment of awkward silence he decided it was time to change the subject. "I wonder why we're still on the ground," he said. "Do they always stay this long when they pick someone up?"

Several of the others smiled or exchanged looks; Jenny Saslov told him, "Mr. Starkman, we took off some time ago, while you were in the shower. This room is completely soundproof, and you can barely even feel the vibration; I guess you didn't notice it over the running water."

"You mean we're moving now?"

Several heads nodded.

He felt foolish. "Oh," he said.

"That's all right," Jenny said. "It's hard to tell."

The conversation languished and died after that. No one had any cards or dice—after all, most of them had been living alone, without much use for such things—and there was nothing in the room except the uniform rows of identical cots and the meager belongings they had brought aboard. There was little for them to talk about, and little urge to talk, since they were all strangers to one another, and all accustomed to solitude. The brief feeling of camaraderie engendered by their mutual predicament had been dispelled, at least temporarily, by the discovery of Starkman's peculiarity and the mixed reactions the discovery provoked.

The only one present whom Starkman felt like talking with was Jenny Saslov; White's reaction to his eyes had aborted any friendship that might have been developing. He found himself, however, unable to think of anything to say; the fact that she was an attractive woman only a few years older than himself, and that he had been so long alone, undoubtedly contributed to that.

Finally, in a state of frustration and incipient despair, he fell asleep.

He was awakened by a general stir in the compart-

ment. He sat up, straightened his sunglasses, and saw
that the large door, the door that he had entered through,
was opening. Two ragged young women, flanked by
Mike and Joey, stood in the passage, fear and wonder
on their faces.

They were ushered in, as he had been, and the door
closed as the round of introductions began.

An hour later a lone man in his late twenties joined
them; some time after that, a family of three was picked
up.

Starkman hung back from the conversations. No one,
so far as he had been able to tell, had told the newcomers
that he was a freak, but most of the group displayed a
visible reserve around him, and the new arrivals picked
up on it, avoiding him without knowing why. He made
no effort to change this state of affairs; he was used to
it. The old hopelessness that had been so much a part
of his childhood was coming back to him after years of
dormancy; like riding a bicycle, it was something one
never wholly forgot.

The cots gradually filled, and the room, large as it
was, began to feel crowded. Food consisted of one meal
a day, but that one generous; it was delivered to an
adjacent room, reached through the door opposite the
"bathroom," through sliding panels in the walls. The
only living things the prisoners saw other than themselves were Mike and Joey delivering more captives,
and occasionally, if a larger than usual group was picked
up, one or two of the other "zombies."

As the number of captives increased, disputes began
to arise over the use of the bathroom, and the door to
the dining room—which was completely unfurnished,
simply a blank metal cube four meters on a side with
sliding panels in three walls—was kept open to provide
more space.

Starkman and a few of the others occasionally ques-
tioned the ship's commander, and Starkman arrived at
the conclusion that it was essentially a disembodied
brain that had been produced somehow by genetic engineering, hooked into an elaborate electronic computer
that provided it with sensory input and additional
memory and computational capacity. The voice insisted

that there were other aspects to it as well, but Starkman could not make sense of its descriptions.

He did not bother to share his theory with the others, or to inquire after their opinions; the subject wasn't important, after all. He did wonder that such a thing had been made; at his last contact with civilization genetic engineers had been making nothing more complex than bacteria. Despite what the zombies had said, and despite the strangeness of the ship's design, he still didn't believe in the Galactic Empire the ship claimed had created it.

He asked as well what lay behind the fourth and final door in the room, and received no answer except that he and his fellow captives were not permitted to use it. That, of course, only increased his curiosity; he began trying to devise some way of prying it open or drilling through it.

There was no distinction between night and day in the windowless chamber; the lights never dimmed or went out. Molley had a wristwatch which he claimed still worked, but he did not care to spend his time telling other people what it said, and would not speak to Starkman at all. Keeping track of time was therefore difficult.

At one point, however, it gradually sank in that it had been a long time since anyone had been added to the party. This was perhaps three or four days after Starkman's capture, and thirty-three of the thirty-six cots were occupied.

He debated mentioning his impression to Jenny Saslov, and had just decided to do so when he felt the faint change in vibration that he had learned meant a landing. He shut up before any sound had emerged, and returned instead to wondering how the fourth door might be breached; he was running through the available tools for the hundredth time when, unexpectedly, the large door slid open.

This was not in accordance with established procedure; it usually took several minutes for the search party to locate and bring back their new captive or captives. The door usually stayed closed for that time. Its opening generally came anywhere from ten minutes to an hour after landing, and in Robert Carvel's case a good three hours had been required; this time, however, the ship had scarcely touched down when it slid aside and revealed, not Mike and Joey and a frightened prisoner, but the full search party of six and nobody else.

"We're here," Joey announced. "Everybody out."

There was a sudden roar of questions and conversation; the ship's men said nothing, but simply gestured for the captives to move out of the room and down the corridor.

Starkman complied as rapidly as anyone; he was tired of being cooped up with a bunch of unfriendly people in such cramped quarters, after spending a decade with all the space he could want.

The immense hatchway was open and the ramp lowered; he followed others down the passage, trying to see over their shoulders what lay beyond. Behind him the remaining captives were doing much the same.

At the top of the ramp, just beyond the hatch, he paused and looked around. To either side were the blank metal walls of the ship's entryway, but directly ahead, between the lower edge of the ramp well and the tops of people's heads, he could see a thin slice of the outside world.

There was no snow.

The first few captives were stepping off the ramp onto bare gray pavement, and throwing long black shadows across it; the sun was obviously shining low in the sky.

He moved down the ramp, and with each step he could see farther. At the second step he saw that the pavement ended in a chain-link fence about thirty meters away, and that there was green vegetation beyond the fence. At the third step he could see trees and green grass. By the fifth step he could, by stooping, see off to either side, and to each side he saw that another ship, identical to the one he was leaving, stood a few tens of meters away. At the eighth step he glimpsed blue sky.

When at last he stepped off the ramp onto the hard gray surface he could not resist stopping for a long, slow look around, much to the annoyance of those behind him. He was far from alone in this, so that it took several minutes for the six zombies to herd their thirty-three charges off the ship.

The vessel was resting on an immense landing field of some smooth gray substance that Starkman couldn't identify; it wasn't quite asphalt, nor concrete, but similar to both. The field stretched off for a kilometer or more to his left, two hundred meters or so to the right, thirty meters ahead and five hundred meters behind. One ship stood to the right, just beginning to disembark its own complement of people, presumably captives like himself and his companions. To the left a row of ships stretched off as far as he could see. Behind stood another row.

The ships were not all identical; there were half a dozen of the same type as the one that had brought him, all lined up together with his in second place, as seen from the fence. Beyond that, and in the other row, there was a wide variety of shapes and sizes; none bore any markings that he could see. Several were discharging people onto the pavement.

The wire fence extended as far as he could see around the edge of the field, which was almost perfectly smooth and level; it was pierced by a wide, open gate at the nearer end, midway between the two rows of ships.

Beyond the fence, in front of him, there was a narrow strip of rough, untended grass; behind that he saw what appeared to him to be primeval forest. Tall, narrow trees stretched up twenty and thirty meters and then ended in puffs of leaves and branches; many looked unhealthy, and the vines that intertwined them appeared to be dead in places. Lower down were trees that seemed healthier and more familiar; he saw a few young oaks that were visibly flourishing, their leaves lushly green. It was a delight to see a live tree other than an evergreen.

Off to the right, at the near end of the field, the forest gave way to a cluster of buildings, a haphazard mix of dull concrete and unpainted wood with a very temporary look to it.

To the left the sun was on the verge of setting and the sky was beginning to take on the rich pinks and yellows of sunset. Elsewhere the sky was a rich pure deep blue such as he had not seen in years, and he stared at it in wonder; he had forgotten so fine a color could exist.

"Where are we?" someone asked.

"This is our main base," Joey replied.

"In Brazil?"

"It's in Brazil, yeah, but it's our main base for the whole planet. Capital's over that way a few kilometers." He pointed toward the buildings.

Carvel was peering at the forest. "I never heard of oak trees in Brazil," he said.

Joey shrugged.

"Hey, you in the ship!" Carvel called.

"Yes, Mr. Carvel?" replied the calm and familiar voice of the ship's brain.

"What're oak trees doing in Brazil?"

"Mr. Carvel, I am not wholly familiar with the ecology of your planet, but I believe that the climatic shift which has recently occurred has altered the local vegetation significantly."

Starkman was not particularly interested in the nature of the local vegetation; he said, "You say there's a capital over that way?"

"Yeah," Joey replied. He seemed to be over his fear of Starkman's fists; his expression now was sullen resentment.

"Capital of what?"

"Of the world, stupid," he was told.

"Is this the capital of the Galactic Empire?"

"Hell, no, that's about a million light years away; this is just the capital of Earth." He noticed that all the captives were off the ship and called, "Okay, everybody, come this way," then began, with the aid of his five companions, herding the group toward the open gate.

Most of the party went along readily enough, Starkman among them; he saw no point in staying on the field. There was no way he could hope to escape by ship; even if some of the ships might not be controlled by built-in brains, as the one that had brought him was, he would not be able to fly one. He hadn't even driven a car in more than ten years.

As the last stragglers were coaxed away from the ship the ramp began lifting up and the hatch at its top slammed shut.

As the group made their way across the pavement Starkman caught up with Joey and asked him another question.

"Are you people from Earth?" He gestured, taking in the other five zombies.

"Of course not, dummy! We're from the Galactic Empire!"

"From the capital of the Empire?"

Joey frowned. "It's none of your business."

"I was just curious," Starkman replied.

"Well, if you've gotta know, I was born on a starship on the way here. It's a long way between stars."

"What about the others?"

Joey shrugged. "Ask them."

"Are all the people of the Empire human?"

Joey stopped walking and looked at him. "Who said we were human? I already owe you a punch in the nose; are you trying to make it worse by insulting me?"

"No, nothing like that—and any time you want to try punching me, you just go right ahead." Starkman was frustrated and angry after the events of the last few days, and would have welcomed an excuse to take out his anger by beating someone. He was not going to start anything himself, for fear of angering whoever was in authority, but he was not going to hold back if Joey started a brawl. He knew he'd be making the man— if he was a man—a scapegoat, taking the blame for more than he deserved, but he didn't much care. Starkman would settle for a scapegoat if he couldn't lash out at the whole world.

When Joey did not take him up on his invitation, he added, "You *look* human."

"Yeah, well, if a five-meter slimy monster had come looking for you, wouldn't you have put up even more of a fight than you did?"

"Are there five-meter slimy monsters running the Galactic Empire?"

Joey clenched a fist, but did not use it. "Look, that's none of your business!"

Starkman started to ask another question, but Joey cut him off. "Just shut up, will you?"

Starkman shut up. Joey's answers weren't much help anyway.

A few seconds later, as he neared the gate, there was a roar behind him; he turned and watched as a small ship, shaped something like an inverted speedboat, lifted off.

Joey shoved him, hard, and ordered, "Keep walking."

Starkman whirled and landed a solid punch squarely in the young man's mouth. Joey went down, landing hard on the pavement, and Starkman walked on toward the gate as two of the other zombies ran to aid their fallen comrade.

Starkman felt much better for a few seconds, but then began regretting his childishness; he would do himself no good, he knew, by attracting attention.

By the time he reached the gate his group had blended with people coming from other ships; all told, Starkman estimated there were several hundred, almost all ragged, carrying a few possessions or none, dressed for cold climates and looking about uncertainly.

Ahead of them he saw where they were being led and/or driven; a line was forming at the entrance to a ramshackle wooden structure which was almost completely covered on its near side with writing in various languages. He guessed that each of the messages said the same thing in a different tongue, since each segment of wall was of a similar size and bore a single short phrase, so far as he could tell. There was no way he could guess at what was said in Burmese, Korean, Japanese, or other languages using symbols totally alien to him, but after a few seconds he spotted a sign in English, located prominently above what appeared to be the main entrance. It read "Welcome Center." Directly below it was the same thing in Spanish, and above it what Starkman guessed to be Portuguese.

The newcomers were being formed up into an orderly column that moved slowly into the building, through the door beneath the English sign.

He had a momentary image of cattle being led to slaughter, but he dismissed it quickly as absurd. A con-

cern he thought worthy of more serious consideration
was that the building was devoted to some sort of brain-
washing, and that he was about to be converted into a
mindless slave—or perhaps a stupid, obedient zombie.
He glanced sideways at the nearest of the peculiar young
men; it was one from another ship, not one of the six
that had picked him up, though there was a strong
resemblance between this one and Mike, as if they might
be brothers.

Would he be one of them soon, thinking himself to
be an alien against all logic?

He looked about and considered making a dash for
the forest, but gave up the idea; there were dozens of
the so-called zombies scattered about, all in their identical blue outfits. They had removed their fleece-lined
coats, which would have been totally inappropriate to
the warm and gentle Brazilian climate—Starkman was
reminded that he was still wearing his own battered
and uncomfortably warm coat—and they were clad in
blue shirts of some synthetic fabric and trousers that
weren't quite jeans. They had retained their black boots.
There was something peculiar about the shirts, and it
took Starkman a moment to realize it was a lack of
collars and pockets.

There was also something unsettling about their
faces; too many of them seemed to resemble each other,
as if all the zombies were members of no more than
eight or nine different families.

He dismissed that from his thoughts and concentrated on his own situation. If he were to create a sufficient diversion, he might be able to get away—but
once he was free, what good would it do, and how long
would it last? He knew nothing about Brazil, what he
might find to eat, where he might find shelter; furthermore, he was sure he could be tracked as easily
here as in Pennsylvania.

He was still inventing and rejecting schemes when
he was propelled out of the dimming sunlight and into
the building.

The line of people was being divided in half, and
each half directed to a door; he was in a lobby or an-

techamber from which only those two doors led farther
into the building.

There did not appear to be any sorting going on, but
simply a division at random, to reduce crowding at the
doorways. When a zombie indicated that Starkman
should take the right-hand door, he refused and bore
left; the man shrugged and let him go.

He noticed that the doors and the proportions of the
antechamber were nothing like the bizarre dimensions
of the ship and its compartments; the doors were completely ordinary double doors, two meters high and each
about three-fourths of a meter wide. The ceiling was
two and a half meters from the floor, and the room was
about four meters deep and twice that in width.

His line inched forward, and he was able to see
through the doorway; he was entering an auditorium,
very much like an ordinary movie theater or lecture
hall. The floor sloped down away from the doors, and
zombies were escorting people to seats, filling up rows
from the back forward. At the far end was a blank white
wall with a small raised platform in front of it, and a
simple wooden lectern stood on the platform. There were
closed doors in the lower corners, as well as the pair at
the back through which the lines were entering. Starkman wondered again whether he was about to be brain-
washed; the blank wall might serve as a screen for some
sort of hypnotic projection, he thought.

Once inside he was led to a seat in the center section,
about halfway down—near the exact center of the hall,
where he had no chance of slipping out unnoticed. To
his right sat the man who had been ahead of him in
line, a total stranger, presumably arrived on another
ship, of medium build and early middle age. He said
nothing. To his left the empty seat was quickly filled
by Ruth Vandeventer; her parents and siblings took
the seats beyond, with Lazarus sitting in his mother's
lap.

He glanced around and spotted several of the others
from his ship scattered about behind him. He looked
for Jenny Saslov and her children and finally found
them off to his right.

The room was only dimly lit, and he had trouble

seeing clearly through his sunglasses, but he did not
remove them.

The hall continued to fill for several minutes, but
finally the last seat was occupied; there were more people waiting to get in, though a rough estimate of the
number of seats convinced Starkman that the room held
three or four hundred. There must have been more ships
delivering them than he had thought, he realized.

The ushers closed the doors, herding the excess people back out into the lobby. As they did, one of the doors
at the front of the hall opened and a tall white-haired
man entered.

Starkman could not have said exactly what the difference was, but he was quite certain that this man was
a normal human being, and not one of the zombies' older
kin. He wore a light gray jacket, black trousers, and a
white shirt, and looked very much like an ordinary
professional man—a doctor, perhaps, or an administrator. The only thing that seemed wrong was the cut
of his clothes, which were not quite what Starkman
expected. Styles had apparently changed in the last
decade.

The man stepped up on the platform and took his
place behind the lectern, while the zombies quietly found
places along the back wall.

"Hello," he said, "I'm Carl Murphy. Is there anyone
here who doesn't speak English?"

To Starkman's vague surprise, there was no answer;
he had not really thought it over, but had assumed that
the people seated on all sides and gathered by a goodly
number of ships had come from widely-scattered places,
and that therefore they would have an assortment of
languages amongst them.

"Let me just double-check, and I ask you all to help,"
the speaker said. "I'm going to ask you to raise your
hands, and anyone who doesn't speak English will presumably not raise his or her hand—or at least, not as
promptly as the rest of you. If any of you spot someone
who is slow in responding, or if anyone knows of someone here who doesn't understand English, I ask that
you point that person out. No harm will come to him
or her, but my little speech here will be given entirely

in English, since all of you come from a region where English is or was the predominant language. There won't be any translation provided, and it won't do any good for someone who doesn't understand what I'm saying. Here at the Welcome Center we have people who speak just about every language known, and anyone here who doesn't understand English will be sent to hear a welcoming speech, like mine, in his or her own tongue, and focusing on his or her own people. Does everyone understand?"

There was a dull murmur of assent; Starkman sat silently.

"Good. Now, everybody raise your right hands."

Hands rose, Starkman's among them; he had no wish to call attention to himself. No one was found who did not understand English.

After a moment, the speaker went on. "Good. You can put down your hands. Thanks very much for co-operating. I realize that some of you may be resentful of the treatment you've received, and it's good to see that you're giving us a chance to explain before making any trouble."

At least, Starkman told himself, whoever was responsible for his abduction wasn't living in a dream world and expecting him to like it. He liked to think of himself as a reasonable person, and was willing to hear the promised explanation before making up his mind whether his kidnappers were insane, dangerous, or merely misguided.

"Now, I'm sure that all of you are wondering why you were taken from your homes, or wherever you were found, and brought here. You probably want to know who the people are who brought you, and what's been going on in the world while you were living in isolation in the wilderness, out of touch with civilization. I'm here to answer all those questions, so far as I can, and any others you may have, as well. What I'm going to do is t give you a brief outline of recent world history, and a description of the current situation, and then explain why we, the representatives of the Government of Earth, have had you brought here. If you still have questions when I'm finished, I'll do my best to answer them. Okay? Now, I hope that you'll all bear with me." He smiled, and touched a button.

Chapter Four

The lights dimmed, and an image appeared on the white wall behind the lecturer; Starkman glanced at the rear of the hall, but saw no sign of a projector. He guessed that there was a rear-projection system of some sort behind the wall, though at first glance it had appeared to be ordinary painted wallboard.

The image was moving; it was film footage of a tropical sunset, presumably from before the current ice age. The sun was an immense red ball in an orange sky, sinking behind green-black foliage. Starkman had a sudden fear of subliminal hypnotic messages.

"As you all know," the lecturer said, "starting early in the last decade of the twentieth century, for reasons that we still do not really understand, significantly less of the sun's light and energy has been reaching the Earth's surface than before. As a result of this change our world has become much colder; where previous ice ages resulted from shifting of the poles or relatively minor drops in temperature over a long period of time, this time average temperatures have dropped fifteen to twenty degrees centigrade all over the world in less than half a century."

The projected image changed, replacing the red sun with falling snow and dying vegetation. Starkman, wary of subliminals, avoided looking at it, for the most part.

"This drop in temperature meant, of course, that the glaciers in Alaska and Canada and the Rockies began spreading in all directions, including southward into inhabited lands. Winters throughout North America became much, much longer and colder, and most of the

continent quickly became almost uninhabitable—more
with every passing year. The same things happened in
Europe and Asia, in Australia, in southern Africa and
the southern part of South America—everywhere, in
fact. Even where the snows could not reach, the temperatures dropped; lands that had never known winter
saw their first frosts, tropical jungles dried out and
began dying. All over the world the climate changed
drastically."

Starkman recognized the film that now appeared from
some of the last news reports he had seen on television
before things collapsed; it showed shivering peasants,
their belongings slung on their backs, camped out in
blowing snow in the shadow of the Colosseum in Rome,
where they had filed when their farms in northern Italy
had frozen over. Ice floating in the canals of Venice,
wind-blown snow whipping past the spires of the Taj
Mahal—the images were so familiar as to seem almost
meaningless. To Starkman the snow and ice seemed far
more normal than the warm sun that had bathed these
landmarks for centuries, more normal than this cozy
lecture hall full of people and illuminated images.

"You know all that, and you know that when people
began to realize what was happening they began moving south to escape the oncoming cold. At first it was
a fairly orderly progress; single families would pack up,
sell their northern homes, and head south, in a steady
growing trickle. Real estate values dropped to nothing
in the north and soared in the south, and the economy
of the world started to show the strain. I'm sure some
of you remember the overpopulation problems that
Florida had in the early nineteen-nineties, and the fuss
over attempts to pass laws against further immigration
from the other states. There were some noble efforts
made—the crash program to irrigate the Sahara, for
example, to provide a homeland for the people fleeing
Europe."

Films of men and machines digging through mountains of sand lit the wall, and Starkman remembered
reading somewhere of a method for spotting some kinds
of subliminals. He wasn't sure where he'd read it, or
how reliable it might be, but it was better than nothing;

he held his fingers horizontally in front of his face and waved them quickly up and down. The idea was that this would create a stroboscopic effect that would let him see single frames in the film, and with a little luck, if subliminals were in use, he would catch one.

He saw nothing but the irrigation project, and stopped when he noticed Ruth, in the next seat, staring at him curiously.

"As conditions worsened, though," Murphy was saying, "the real panic began. The United States and Canada collapsed as their populations fled southward and their governments failed utterly to cope with the crisis. A hundred gimcrack dictators tried to fill the void; maybe some of you were first driven into isolation by those attempts. Communications fell apart, and that probably isolated the rest of you. The entire United States lost contact with the new mainstream of civilization around 1998, when the Constitution was formally suspended and even some of the petty dictators gave up and fled south. I'm sure that most of you haven't had any contact with any sort of government since then."

This speech was accompanied by scenes of desperate people mobbing planes, overloaded boats sailing clumsily out of port, uniformed figures haranguing crowds, and empty, snow-covered city streets, drifts climbing up storefronts under the pressure of the wind. Starkman surreptitiously waved his fingers again, but still found nothing suspicious.

"Well, it was late in 1999 that our friends from the Galactic Empire arrived—almost twelve years ago now. An exploratory fleet sent out by the Galactic Empire discovered Earth at that critical juncture, and stepped in to aid their fellow sentient beings."

The scene now was unfamiliar, a vast array of ships setting down one after another on what appeared to be an ordinary airport, filling the runways and taxiways and overflowing onto the surrounding fields and streets. The image shifted to several men resplendent in gaudy uniforms walking down a ship's ramp, hands held up in a gesture of peaceful greeting. Starkman thought they might be zombies, though their beards appeared to have been trimmed; they had a sort of vagueness to their gestures, as if they were following rote learning rather than acting of their own will, that reminded him of the search parties.

"I'm sure a lot of you are wondering what the Galactic Empire is, and what motives impelled its people to intervene. Well, I don't know too much about the sociological details—no one on Earth does—there are problems in communication that haven't yet been cleared up. Apparently, however, there is a vast federation of intelligent beings that dominates a large part of our arm of the galaxy. 'Galactic Empire' is a very rough translation of the right name for this, a feeble approximation we use for convenience; there is no all-powerful emperor, no conquering armies, as the term 'Empire' might seem to imply. Inclusion in the Empire is entirely on a voluntary basis, and it is a highly cooperative and democratic organization. The people of many worlds are included, on an equal footing, and no one race or species rules; their capital planet was chosen for its neutral location, and was previously uninhabited. With the resources of so many worlds and peoples at its command, the Empire has eliminated almost every social ill on all their member planets—and upon arriving here and finding our world in such a mess, the exploratory fleet's commander resolved to help us cure our own problems, though we are not yet sufficiently advanced to become a partner in the Empire."

Throughout this part of the speech the wall showed an alternation of the smiling uniformed zombies and pretty, meaningless pictures of starry skies and colorful nebulae; Starkman was disappointed that there were no scenes from other worlds. He would not necessarily have believed them in any case, since the whole idea of a Galactic Empire still seemed far-fetched and unlikely to him, but he regretted that no one had even come up with interesting fakes.

"Accordingly, several ships landed,"—the scene returned to the airport crowded with starships—"and representatives who could speak our various languages were sent to parley with the surviving governments of the world." A series of scenes of uniformed zombies

speaking to official-looking audiences, or facing impor-
tant-looking people across conference tables, began.

"Most governments accepted the offer of aid immediately, and turned the reins of power over to the
Governor, the exploratory fleet's commanding officer,
a special administrator trained in organizing, who could
be relied upon to manage affairs here on Earth as best
they could be managed. Those who did *not* accept aid
were ignored at first, left entirely to their own devices;
those of us who were cooperating were far too busy
rebuilding our societies to bother with them. With time,
as they saw how successful our alien benefactors were
in restoring order and prosperity, most came over gladly.
In some cases where governments would not cooperate,
their citizens deserted them en masse to join us. The
last holdouts were finally forced to cooperate with
threats and coercive measures. Only in a handful of
instances was there any resort to violence to bring individuals to cooperate—and let me emphasize very
strongly that such violence as did occur was always
instigated by Earth people, against Earth people. The
Governor and his people took no part in it, and in fact
voiced strong disapproval, saying that with patience
everyone would come around in time. The fleet has
made plain that it will not interfere in our internal
affairs uninvited, and that the Galactic Empire does
not want harm to come to anyone. Throughout the es-
tablishment of the Governor's authority here all extraterrestrials were very insistent that no one should be
harmed if it could possibly be avoided. For those of you
who fear that the Empire intends to conquer Earth, let
me say that there is no recorded instance of a citizen
of the Galactic Empire, or any of their machines, harming any human being."

Starkman remembered the boiling lines of snow and
snorted quietly in derision. He did not think the machine controlling the ship had been bluffing. He
suspected that it was a matter of careful record-
keeping than unbounded beneficence that allowed the
lecturer to make such a statement.

"Furthermore, out of respect for our sovereignty, and
to allay our natural concern, the Empire has made no

attempt, despite their pacifist beliefs, to destroy or con-
fiscate any weapons, so that mankind still has available, should they be wanted, arsenals as complete as
before the fleet's arrival. Everything from handguns to
nuclear submarines and hydrogen bombs is still stock-
piled, in the possession of Earth people. We still have
our own spacecraft, primitive though they may be; un-
fortunately, the fleet does not have the authority to
teach us the secrets of their interstellar drive. That
information is, of course, carefully restricted, to prevent
barbarians from disrupting the peace of the galaxy. It
may be that when the word of our existence reaches
Imperial officials and an evaluation is made, we will
be given those secrets."

Toward the end of this portion of the speech the pictures of government councils were replaced by old footage of warships and spacecraft; Starkman thought that
some of the scratchy films dated back to the Reagan
administration or even further, and he was quite certain that they were no evidence at all that the arsenals
Murphy mentioned actually existed.

"The fleet *has,* however, provided whatever help its
people were able, under their own laws, to give. They
have helped us to build cities that float on the sea and
harvest plankton for food, using the tides and wind for
power; they have helped us put colonies on the ocean
floor, as well."

The images of the floating cities were new to Starkman, and quite impressive; there were signs of use and
an air of solidity about them that he had never seen in
special effects, and though still cynical and cautious,
he was willing to believe they were real.

"With the aid of genetically engineered crops and
food animals, we and they, working together, have made
the seas and wastelands flourish and yield, solving completely the critical food shortages that resulted from
the loss of so much of the world's arable land to the
snow and ice. The Sahara, which we had just begun to
conquer, is a paradise now. The steepest mountains can
be farmed."

More very impressive scenes accompanied these

statements, so impressive that Starkman wished they weren't changed so quickly.

"They have put their ships and machines at our disposal, using starships to transport people and matériel wherever they are needed, or wherever they might wish to go, allowing us to redistribute the world's population for greater efficiency, so that all can live in comfort. It is only due to the advanced technology of the Empire that mankind can survive and provide enough food for all."

Street scenes of smiling people seemed somehow the least convincing images yet; Starkman remembered the crowded streets of his youth, and never had he seen so high a percentage of smiling faces.

"Now, some of you may say that this has interfered with the natural order of things, and that in moving our population into the deserts and mountains and oceans we have endangered or destroyed the world's ecology. I do not want to dismiss this charge lightly; it's a legitimate concern. However, there are important extenuating circumstances that must be taken into account. Firstly, the world's ecology was already breaking down rapidly under the disastrous effects of the new ice age; nature herself had destroyed the balance that had existed for as long as our species has walked the Earth. Simply look at the forests outside this building and you will see clear proof of what I say; we are in the heart of what was once the Amazon rain forest, and it was no act of man that destroyed the tropical jungle, but simply the change in climate. Secondly, we had little choice; it was either a complete restructuring of our society and the exploitation of all the wilderness still warm enough to use, or the deaths of millions upon millions of human beings like ourselves. It was only the arrival of the fleet from the Empire that gave us any choice at all and saved those millions of lives."

And wasn't that a lucky coincidence, Starkman thought as the pictures alternated between familiar scenes of frozen desolation and images of the new cities and farms, contrasting forests dying wrapped in ice with lush green crops. He was sure that there had to be some correlation between the ice age and the emergence of this "Galactic Empire"; he guessed that the whole Empire was run by a conspiracy of scientists who had decided to exploit the natural catastrophe, to use it as an excuse to seize power. If there really were any aliens—he was still convinced that aliens would not be so perfectly humanoid in appearance as the zombies, and dismissed as ridiculous their claim to be extraterrestrial in origin—if there really were any aliens, they had probably come when they did not by coincidence, but because the ice age provided too good an opportunity to pass up for whatever their purpose might be.

Murphy was continuing his speech.

"And not only have all those lives been saved, my friends, but life is now better than ever in many ways, thanks to the knowledge brought to us from the stars! Through biological engineering far beyond our own feeble attempts there is food enough for everyone, and better, safer, more nourishing food than ever before."

That claim worried Starkman with the thought that perhaps his meals aboard the ship had been drugged, and that he might already have been affected somehow, already compromised.

"Knowledge of the basic principles of biology, the laws of life, has permitted the people of the Empire, though different from ourselves, to improve the state of human medicine until virtually every disease known to our long-suffering species can now be completely cured! Yes, I include cancer in all its terrible forms!"

Starkman watched people in hospital gowns go from sickly, emaciated, bedridden invalids to healthy, happy, smiling individuals; superimposed dates indicated that these transformations took no more than a few days or weeks.

"Some of you may fear that all these benefits, this greatly increased health and security, may have come at the cost of our freedoms."

Starkman almost nodded in response; the speaker had described his suspicions perfectly—with the additional item that he did not yet wholly believe in the supposed miracles of alien science. He was not convinced that the "Galactic Empire" existed anywhere other than Earth.

"Fear not! The aliens are acting out of the highest moral motives, and have left our ways of life untouched. We still enjoy as much freedom as we ever did, freedom of movement, freedom of belief, freedom of choice in every field. Some of you may have stayed behind in the States for fear of religious or racial persecution; there was a good deal of such unpleasantness during the days of chaos, as various groups fought to survive. Now that there is peace and plenty, however, and all the world united under a single benevolent government, there is no reason for such persecution, no need for man to compete against his brothers for the world's resources. Every effort is being made to assure that everyone shares equally in our good fortune."

Starkman thought that the scenes of racially mixed church choirs and black men shaking hands with white were somewhat oversentimental.

"There are still, of course, flaws in our paradise, and I won't hide them from you; you would then be telling yourselves that this is all far too good to be true."

Once again, Starkman almost nodded agreement.

"There is still crime—but less than before, since no one goes hungry and we have better techniques than ever before for dealing with criminals. There is still intolerance in all its forms—but a vigorous public education program is under way, and various legal protections are in effect. Progress is being made. I regret to say that there are certain temporary minor restraints on free speech and freedom of the press; although nothing has been totally suppressed, the government has required that certain books and pamphlets carry disclaimers stating that their content is not accurate or not in accord with official policy. You will quite probably see some of these pamphlets in the near future; they are regularly distributed to new arrivals in our city of Capital by a small group of the malcontents that are inevitable in any society."

The image of a crudely printed pamphlet entitled *Resist the Alien Conquerors!* fascinated Starkman; the fact that it was shown here convinced him, more than anything else, that the Galactic Empire's Government of Earth was either extremely subtle, or else just what it said it was.

Or, possibly, it was both, and the important things were what it did *not* say, either to affirm or deny.

The red rubber-stamped message on the pamphlet read, "This work makes inaccurate claims and states positions considered unsound." The incredibly mild tone of the government's statement made a strong contrast with the huge smeared letters of *Resist the Alien Conquerors!*—a contrast that was surely intentional. Someone in the government knew what he or she was doing, playing the voice of calm reason confronted by raving fanatics. He wondered whether perhaps the government had printed the pamphlets itself, to play up that idea.

"Now, you may be saying to yourselves, 'This is all very interesting, but what does it have to do with *me?* Why was I taken from my home and brought here?'" The speaker smiled broadly. "Well, it's really very simple. When someone has something he thinks is a good thing, a better way of life, his impulse is to share it with others. Religious evangelists, for example, feeling that they have the one true way, try to teach it to others. The United States, in its vanished heyday, tried to export its political system and free economy; before that, the British Empire spread English customs around the world. Why did they do this? Simply because they thought that they had found the right way to live, and they wanted the whole world to share in their good fortune."

Scenes of Peace Corps volunteers of the nineteen-sixties lecturing African villagers filled the screen. Starkman thought that was less than apt; Americans would not take kindly to being identified with what they thought of as primitives.

"Similarly, those of us fortunate enough to be living under the new world government think that we have found a better way to live. We may be wrong; our way may be right for us, but not for you, just as British or American ways were not right for everyone. However, we did want to see that everyone on this big blue globe had the opportunity to decide for himself or herself. We

wanted each of you to know what we had, not from
vague rumors that might drift up by word of mouth,
but firsthand. We wanted to give you the blessings of
good health and plentiful food that we now enjoy. We
didn't want anyone, anywhere, to die of starvation or
neglect because we couldn't spare the time to find him
or her, living alone in the snowy wastelands that were
once home to most of the world's population. The Galactic Empire brought us peace and plenty out of the
sheer goodness of their hearts—or whatever organs our
benefactors may consider the seat of the emotions. Could
we do any less for our own brothers and sisters, members of our own species? With that in mind, we asked
the citizens of the Empire who were not occupied with
more urgent tasks to help us in helping you, and they
very generously complied. The fleet sent ships and crews
to search the cold wilderness for whatever survivors
might be found, to bring them here so that they might
see what opportunities await them."

Thinking it over, Starkman thought he might have
been too hasty; watching scenes of zombies gathering
in people, as he had been gathered, he wondered whether
resentment of being identified with the Africans might
impel people to prove that they were *not* primitives by
giving up the precarious life of the wastes for the civilized comforts of the Empire's world. It was a subtle
point, but he was sure that it was not beyond these
people.

He wondered how the inept, childish zombies could
be connected with the people who planned this speech.
They were obviously not the same people. The zombies
must be, he decided, a servant class, perhaps bred for
the purpose. They themselves might not even be aware
of their true status.

"What does await you here? Well, firstly, you'll all
be given quick medical examinations, and any disease
or other medical problem that you may have will be
treated as best we and the doctors of the Galactic Empire can contrive to treat it—and in almost every case,
that means that you won't have any more medical problem. Afterward many of you will probably be healthier
than you have ever been before. Those of you who are

willing will then be interviewed; we're curious about conditions in the northern lands, and will appreciate any information you can give us. It's strictly voluntary; you can skip that if you like. If you do answer questions, we'll return the favor and provide any information we can, and answer any specific questions you might come up with. After that, you're free to do as you please; we've got apartments waiting for you for tonight, and in the morning, if you like, we'll be glad to return you to wherever we found you, in the wastes of Pennsylvania and Ohio and West Virginia—does that cover this whole group? Oh, Maryland, too. Wherever you might have been picked up, we will return you to that exact spot when we send the ships back to look for more survivors."

Starkman wondered how authentic that promise was; had anyone ever taken them up on it? He was unsure whether he would himself, he had to admit; he had no great love for cold and hunger.

"If you choose to stay here, in the warm countries, under the new world government, then we'll find a place for you, a new home and a good job. If you have relatives who might still be alive and wondering about you, we will try and put you in touch with them. Even if you plan to return to the States, you might want to let your relatives know that you're still alive. We have a computer file on just about everybody in the comnet—the computer and communication network, that is. It's nothing to be frightened about; don't be concerned for your privacy. It's just the medical records from our worldwide health programs, with cross-referencing, but we can probably locate any family you may have."

That offer didn't much interest him; the only person who might care about him was a younger sister he'd never gotten along with.

"I think that's just about all I planned to say. Now, if any of you have questions you think may be of general interest, I'll try to answer them; raise your hands, and I'll try to call on you all. If you have questions of a private nature, questions that probably wouldn't interest most people, please keep them until later and ask the doctors or the interviewers, rather than taking up everybody's time here. Now, who has a question?"

The wall behind the lectern went blank, and lights came on around the room. A dozen hands sprang up; the lecturer pointed to one in the front row.

A scrawny young man stood and demanded, "Who're you? Why are you telling us all this stuff?"

"As I said, my name's Carl Murphy," the speaker replied, "and I was asked to give this speech by the government agency in charge of collecting stragglers. I was an assistant professor of psychology at Princeton up until 1996, when I decided I'd rather live someplace warmer. I spent three years wandering about in Central America, and then was recruited by the new government to work for their public information service. I'm telling you all this stuff so you'll know where you stand, and not be confused or lost or unreasonably angry about being brought here. We didn't want to just dump you all into Capital unprepared; that'd be cruel, after snatching you out of isolation, to dump you in the middle of an unfamiliar civilization."

The youth seemed unwilling to accept this, but someone beside him pulled him back down into his seat; Murphy watched for a moment, to be sure he wasn't going to protest, then pointed to another raised hand.

A middle-aged woman rose and demanded, "If these wonderful alien benefactors of yours are so damn smart and so advanced, why can't they just stop the ice age and make the world warm again, or at least help people live where it's cold? You're letting most of the world go to waste!"

"Well," Murphy said, "it would certainly seem reasonable to ask that. The fact is, we're working on it. We don't quite understand why the climate changed, so we don't know how to fix it. A simple and drastic solution would be to move the whole planet to a closer orbit around the sun, but that's very dangerous, and the fleet here—which is, after all, just an exploratory fleet—is not equipped for it. Artificial heat could be provided, perhaps—but, once again, this fleet isn't equipped for it. They didn't come here expecting to be asked to fix an ice age. And it's worth considering that if they *were* to provide artificial heat, and then the ice age were to end naturally, we might all fry before the heat could be removed. No, it's just not worth the risks to try meddling with the climate before we know exactly what we're doing."

"But why can't they make the north habitable again?" someone demanded. "They could build domes or something!"

"That's true, they could. They're working on it, and so are we; right now domes are being built over Atlanta, Las Vegas, and other cities around the world. It takes time, though, a great deal of time, and such shelters are very energy-inefficient compared to floating cities or undersea colonies. For one thing, the ocean colonies can harvest their own food, draw energy from the sun and wind and tide, and generally maintain themselves with very little outside support, while Atlanta will have to import all its food and rely on fusion reactors for power—and I'm sure you all know how expensive fusion reactors are to build and maintain. Still, we really are working on it; we aren't planning to let most of the planet go to waste."

The next questioner wanted to know where the fleet had come from.

"If you mean where its crews came from, they came from a dozen different worlds all over the Empire, as did the ships. Its last port of call, before adventuring out into unexplored space, was an outpost with an unpronounceable name circling a star somewhere in Sagittarius. I don't know exactly which star, and I'm not sure it can be seen from Earth at all. I certainly don't know its English name, if it has one."

"Do all the aliens look human?"

That was a question that interested Starkman.

"This answer may surprise a lot of you, but I honestly don't know for sure. All I've ever seen have looked just as human as you or I, but no one has ever seen the Governor himself, so far as I know. It may be that they don't want to frighten or upset us with creatures that look horrible to us, or it may be that some of them can't breathe our air. I'm sure you all think it's very strange that we could have lived with these people for a dozen years without even being able to answer such a simple question, but it's the truth. They keep to themselves, if there are any others besides the humanoids like the ones that brought you here. We respect their privacy. They have intelligent, self-aware machines, which I suppose you could count as non-humanoid aliens, but whether there are any that were born rather than built, we simply don't know."

Starkman concluded that Murphy was answering questions honestly; why else would he make such a damaging admission?

Then he reconsidered. What if Murphy were making such an admission simply to convince everyone of his honesty? It seemed more than strange, almost unbelievable, that he would not know, but professing ignorance would certainly bolster his believability on questions he *did* answer.

He reconsidered again. He might well be attributing more subtlety to the aliens and their flunkies than they actually possessed, assuming that there really were any aliens.

There were no more questions of any interest, and when a minute or so had passed with no new hands raised, Murphy signaled to the zombies at the back of the room. They came forward, and as they moved down the aisles Murphy announced, "That's it; you'll be escorted from here to the medical-exam area. Thanks for listening." He waved, stepped down from the lectern, and departed through the door to the left of the front of the hall.

The zombies took over when he had gone, motioning for the people in the front row to exit through the doors at the front of the hall. With some jostling and protests, they did, followed by the second row, and the third.

When Starkman's turn came he went along quietly; he saw no real chance to escape at this point. He was also still thinking over the lecture, trying to decide how much of it he could believe. He wondered whether there had actually been subliminal images used, and if so whether they might have affected him, perhaps making him more passive.

At that thought, to prove to himself that he had not

been rendered docile, he stuck out a foot and tripped a passing zombie; the young man caught himself on someone's arm and managed to keep from falling, but created something of a stir.

The man turned to glare at Starkman.

"Sorry," he replied with a smile.

The door that Starkman's half of the auditorium was sent through led into a rather shabby white-painted corridor lined with more doors. The doors were all closed, and the people leaving the auditorium were standing in an uneven row down the center of the passage. Every so often a door would open and someone inside would call, "Next!" When that happened, if the nearest human did not immediately step forward a pair of zombies would grab someone and shove him through the door. Whether the subject went willingly or not, in every case the door closed immediately, and that person did not emerge again. Starkman wondered whether the people were accumulating in the various rooms along the hallway, or leaving by some other, unseen route.

The image of cattle in a slaughterhouse came to him again, and he dismissed it. Why would anyone waste time and energy lecturing people going to slaughter?

He would have preferred to hang back, to give himself time to think, but he was not readily able to do so, as more people were being herded out of the auditorium behind him, forcing him forward. He spotted several of his erstwhile companions on the flight south.

Jenny Saslov and her two children wound up standing nearby; he made his way around a pair of old men whispering to one another, moving close enough to speak to her.

She made a vague noise of greeting upon recognizing him; he replied with a mumbled hello, and after a moment of uncomfortable silence said, "What d'you think they're doing?"

She shrugged. "I guess this is where we get our medical exams."

"What's a medical zam?" asked Charlie.

"That's when a doctor looks you all over very carefully to see if there's anything wrong with you—whether you're sick, or haven't been eating right, or have hurt yourself somehow," Jenny told him.

"I'm not sick!" her son retorted, obviously very offended.

"I know you're not, but some of these other people may be, and to make it fair they're checking everybody," she explained.

"Oh." Charlie mulled that over.

"He's never seen a doctor before," Jenny told Starkman.

"I wish I hadn't," he replied.

She started to smile, then stopped. "Oh. I guess that's right."

"They always treat me like a damned specimen, rather than a human patient."

Her face grew worried. "I hope they don't give you too bad a time here."

He grunted.

"Do you think they're really going to let us go where we please?"

He shrugged. "I don't know. Probably; they seem too smart to make such an obvious lie."

"Did you believe that lecture?"

"Not all of it. I'm not sure what I should believe and what I shouldn't; I suppose we'll just have to wait and see."

"I suppose so," she said.

The nearest door opened, and a female voice called, "Next!" When no one moved immediately, a nearby zombie looked around, then pointed to Kathy.

"You next, kid," he said.

Kathy stood where she was, looking at her mother.

"You go along," Jenny said. "The doctor will look after you. Charlie will go next, and I'll be right behind him, and after we've all been examined we'll get back together again."

Kathy still appeared reluctant, but when the zombie reached out for her arm she shook him off and marched through the doorway. Jenny, Charlie, and Starkman watched her go.

The door closed. Starkman tried to think of something to say, to renew the conversation, but nothing came.

After a few moments—a very few moments, Starkman realized, much less time than he would have expected a thorough checkup to take—the door opened again, and Charlie vanished through it.

Jenny followed in her turn, leaving Starkman next to the door and the obviously bored and fidgety zombie. He decided to be the next one in, in hopes of staying near the Saslovs, who were the closest thing to friends he had in the world, or was likely to find.

The door opened, the voice called "Next!", and Starkman stepped forward.

Chapter Five

The examination room was small, less than three meters on a side. There was a privacy screen to the left, marking off the nearest corner, and a closed door in the center of the far wall. The entire right side of the room was jammed full of machinery, none of which Starkman could identify with any certainty. There was what appeared to be a computer keyboard and display, and at least one device had a pen that graphed something or other onto a moving sheet of paper. Several dials and displays were unlit and motionless.

The only other person in the room was presumably the doctor, a short dark woman in her thirties wearing a spotless white lab coat and with a stethoscope around her neck. Starkman wondered whether doctors still used stethoscopes, or just wore them as a badge of status.

"Hello," she said, "I'm Dr. Curtis."

Starkman nodded politely, but said nothing.

Visibly annoyed, the doctor asked, "What's your name?" She waved at the banks of machinery. "We like to keep complete medical records."

"I'm John Starkman," he replied.

"Good. Mr. Starkman, this shouldn't take very long; most of what we want to check will be done by analyzing a blood sample. I'll want to take a look at your eyes and ears and throat, and maybe listen to your heart; after that I'll check out any specific complaints you may have, aches or whatever, then take the blood sample and let you go. There are buses out back to take you to the temporary housing. Good enough?"

"Fine," Starkman said with a shrug.

"Good. Then get over here where I can reach you and where the computer can see, and take off those silly sunglasses. It must be dark outside by now, anyway; you won't need them."

He had been dreading this moment. He started to protest that he didn't need to have his eyes checked, but changed his mind. It would be undignified to make a fuss, and he was sure that he wasn't going to convince the doctor to change her routine without any explanation. Reluctantly, he removed the glasses and put them in his coat pocket.

The doctor had glanced over at the machinery; when she glanced back and saw his eyes, she froze.

He waited for the inevitable questions about what was wrong with him, but they didn't come; instead, she said, slowly and loudly, "Max, cancel the record of this patient's name; we got it wrong. And please give me complete privacy. There's a personal matter I want to take care of."

"Yes, doctor," answered a voice from the bank of machinery.

"Max, I want privacy for us both."

There was no answer. She cast a suspicious look at the computer display, then turned back to Starkman.

"Look," she said urgently, "just do what I say without any questions. Put your glasses back on, and keep them on. Don't let anybody see your eyes. Now, I want you to go back into the waiting line, and wait until everybody else has gone through, then come back here, to me. Don't let any other doctor see you. This is *important*. Don't let *anybody* see your eyes."

Starkman was utterly confused. "What about the zombies? They won't let me back in line, will they?" he asked as he restored the sunglasses to their accustomed place.

"You mean the Imperials? They will if I tell them to."

Starkman wanted to ask further questions, but Dr. Curtis was already dragging him back to the door he had entered through. Before he could decide what to say she had opened it and was gesturing to the nearest of the blue-clad young men.

"Listen, stupid," she said, "this man's a special case. I want him to come last, so I won't have to rush him, okay?"

The zombie looked puzzled. "He's not supposed to come back this way."

"Look, it's all right, okay? I told him to, and I'm telling you to let him. I haven't examined him yet; I want him to be last. Understand?"

The man still looked doubtful, but he said, at last, "Well, okay, if you say so."

"I say so. Next!"

Another captive was shoved forward as Starkman made his way back up the corridor to a quiet corner near the auditorium door.

The crowd in the passageway slowly thinned as, one by one, the captured people were called into the examining room. Starkman hung back, as he had been told, trying to ignore the attention the zombies—or Imperials, as Dr. Curtis called them—were giving him.

Finally the last of the others vanished through a doorway, leaving him alone with two dozen Imperials.

A door opened, and a white-haired man leaned out into the corridor. "Is that it?" he asked.

"There's one more," one of the zombies said.

"I'm waiting for Dr. Curtis," Starkman called from his corner.

The doctor turned to look at him. "I can take you now, if you like."

"That's all right, I'll wait."

The doctor shrugged. "Suit yourself," he said as he withdrew and closed the door.

A moment later Dr. Curtis's door opened; she leaned out and beckoned to him. He hurried into the examination room; she said as she closed the door, "I was afraid the next batch would arrive before I got to you."

He said nothing.

"They aren't due for another fifteen minutes, but sometimes Carl rushes his talk. That gives us maybe ten minutes to get out of here." She crossed to the computer keyboard, punched a button, and spoke into a grille. "Max, I'm through for the night, and the last straggler has gone through; send the buses on without me, I'm going to walk home. Get somebody else in here for the last group if you need to; I've had it. Got that?"

"Yes, doctor," said the machinery.

"If anyone asks for me, I'm not available, not to be bothered."

"Yes, doctor."

"Good. Now, I want you to erase all visual records from today, and then shut off this terminal for the night."

"Yes, doctor." There was a faint click.

Curtis crossed to the other door, opened it, and leaned out. Starkman saw that it led into another corridor paralleling the one he had waited in for so long. After looking both ways, and with a final glance back at the computer, Curtis stepped out into the passage and beckoned to Starkman.

"Come on, quickly," she ordered.

Starkman hesitated. "Why? What're you doing?"

"I'll explain later; right now we've got to hurry. Come *on!*"

Reluctantly, he obeyed, and the doctor led the way down the corridor, through a door marked AUTHORIZED PERSONNEL ONLY in six languages, and into a small, clean locker room. There she removed her lab coat and flung it aside, then opened and rummaged through one of the lockers.

The coat landed in a heap in a corner. Underneath she wore jeans and a light-tan sweater.

She emerged from the locker with a large handbag and a flashlight, then paused, frowning.

"Is there anything else I need?" she asked.

"How should I know?" Starkman replied.

"I haven't got time. Come on." Moving quickly, she led the way out another door, down a shor⁺ passageway, and out into semitropical moonlight.

Seeing the great copper orb in the sky, Starkman wondered what a tropical moon had looked like before the climate changed. Had the lower temperatures affected its appearance?

He didn't know, and it didn't matter.

They were on a sidewalk; a street ran in front of them, along the side of the Welcome Center from the gate of the spaceport and off into the distance in the direction Starkman believed led to the city of Capital. It was lined with street lights that competed with the moon in yellow warmth. On the other side of the roadway from where they stood he could see nothing but forest.

Dr. Curtis marched directly across the street and into the forest.

Starkman followed, hesitantly, and found that she was standing at the end of a path that vanished amidst the trees. It curved back upon itself and ended behind a clump of bushes, so as to be virtually invisible from the sidewalk or most of the road.

Behind him he heard the sound of engines, purring smoothly; he turned back and saw a sleek bus pulling out of a lot behind the Welcome Center onto the road, bound for Capital.

"You were supposed to be on that," Curtis whispered. "Now come on." She clicked on the flashlight and led the way into the forest.

The trail wound through the trees and brush in an irregular serpentine, and Dr. Curtis hurried along it with the ease of total familiarity, hardly even using her light. Starkman struggled to keep up, ducking low-hanging branches that loomed black against the moon, missing unexpected turns, and stumbling over roots and rocks that were invisible in the darkness.

He noticed the calling of nightbirds and the sounds of other creatures going about their own business in the forest around him, as if it were seething with life. In Pennsylvania there had been nothing to hear in the night except the wind and the hissing of snow drifting beneath the cold dead stars, and a moon that was as white and harsh as the one above him now was golden and welcoming.

There had once been birds in Pennsylvania, of course, but they had disappeared so gradually, over a period of years, that he had never noted their absence in the years since they vanished for good.

Starkman was so busy watching the path and listening to the animal noises that he didn't see the buildings until just before he emerged from the forest into the clearing that surrounded them.

There were three of them, immense ramshackle structures of plywood and corrugated metal. Each stood three or four stories high, and was easily forty meters on a side, probably more. They were weatherbeaten and, even by moonlight, in obvious disrepair; grass and weeds grew knee high between the strips of pavement that twined around them and led off in various directions. The forest of trees was mirrored in a forest of old lamp standards, but the lights were mostly dark, and the few that still worked at all managed no more than a feeble fluorescent flickering.

There was a sudden loud rumble from off to his left, and Starkman realized that he was still quite close to the landing field, where a ship was lifting off. He could see the glow of bright white lights through the trees, but could make out no detail beyond the mere fact of their presence.

Dr. Curtis paid no attention to the ship or the lights, but led him toward the nearest of the three buildings. A sign over the door, faded from weather and hard to make out clearly in the semi-darkness, said "Welcome" in a dozen languages.

At the threshold Starkman stopped.

"What is this place?" he demanded. "Why did you bring me here?"

Curtis stopped where she was, halfway down a short unlit passage; she turned and said, "Come inside and I'll tell you."

Starkman took two steps forward into the gloom. "I'm inside," he announced.

Curtis sighed. "Wouldn't you rather come in and be comfortable?"

"Not until I know where I am and what's going on."

"You're in the old processing center; it's been abandoned for about two years now, so we're using it as a meeting place."

"Who's using it? Why was this place abandoned?"

Curtis stared at him for a moment. "Look, I'm tired," she said at last. "Can't we go inside and sit down to talk?"

"Sit on the floor; I'm not tired."

Wearily, she did as he suggested, leaning back against

one wall and sliding down into a sitting position. Starkman watched silently. He noticed for the first time that light was seeping through into the passage from somewhere inside the building, leaking around the edges of a pair of fire doors and pouring thinly through two small windows in the doors, as if there were lights two rooms away.

She looked up at him. "Tired or not, you might want to get comfortable."

"I'm fine."

"Okay. Look, Mr. Starkman, you heard that phony lecture Carl gave, right?"

"I heard the lecture."

"Did you believe it?"

"Not all of it."

"Well, I don't know what you believed and what you didn't, but it's all a clever bunch of lies mixed up with half-truths. Some of us realized that, and we've formed an underground organization dedicated to driving these filthy aliens off our planet. Are you interested?"

Cautiously, he replied, "I might be."

"I brought you here because I thought you could help us."

"I might. Keep talking."

"I don't know what else to tell you."

"You could start by telling me what these buildings were for, and why they've been abandoned when the warm countries are supposed to be hopelessly over-populated."

"Oh, hell, the world's not *that* crowded; nobody wants to live next to the landing field and put up with the noise. I suppose they might want to farm the land here eventually, but so far it hasn't been worth the trouble of clearing out what's left of the jungle. It's easier to use the oceans and deserts."

As if to confirm what she said, Starkman's next question was drowned out by the roar of another ship. When it had passed he repeated, "What were the buildings for?"

"Processing centers, same as the one you came through. They don't need them anywhere this big any-more, since there are only a few of you diehards left to do, so they built the little one and left these to rot. We decided they'd make a handy meeting place."

"What were they processing? Did they find that many people up north?"

"Who said anything about up north? Mr. Starkman, they put every damn human being they could find through their damn processing. So far as I know, except for the crowd they've got in that lecture hall right now, you're the only human being in Brazil they *haven't* processed."

"I don't understand."

"Look, why don't you sit down?"

Reluctantly, Starkman complied, seating himself on the floor a meter or two away from her.

"That's better. Now, one thing that's mentioned in that lecture Carl gives six times a day is that they have medical records on everybody, right?"

"Right."

"How do you think they got those records? When they first arrived, before they did anything else, they had these centers built, not just here, but all over the world. Before they started farming deserts or building floating cities or any of that stuff, they built these centers, and they started putting people through them. They heard that lecture, or one like it; you got the updated version for North American survivors. It was originally a lot shorter and simpler, just saying that they were here to save us. Then every person who had sat through the talk got a medical exam, and blood samples were taken and sent aboard the ships. When that was over, everybody got marked so that the government they were setting up could tell who they'd done and who they hadn't." She held out her right hand. "You can't see it without one of their special readers, but I've got a serial number tattooed on the back of my hand here, and so does everyone else they've processed. At least, everyone who's *got* a right hand has it there; the rest get it on the forehead. It's part of my job to put it there; they've got it all mechanized so it takes about two seconds and doesn't hurt." She smiled wryly. "There are a few of the more religious folk who say that all the serial numbers are the same, just three digits—six-sixty-six. I don't think that's true, myself, but maybe it might as well be."

Starkman sat and considered this for a moment, puzzled. "I don't get it," he said. "Why are they doing that? What do they get out of it?"

"We don't know. They say they just want to be sure they give everyone a checkup, to prevent any nasty diseases from slipping through undetected."

"Maybe that's the truth."

"Maybe it is, but we don't like it, whether that's the truth or not. What it means is that they've got everyone on file, everyone numbered. You can't enter a government office anywhere, for anything, without being checked to see that you've got a number. You can't get a job most places without it. You can't get to their new cities. You can't use their transport systems. That's the wonderful freedom they promised us."

"But all they do is give a lecture and a checkup, and then you get your number?"

"That's about it, yeah. But don't forget the blood samples. They won't let us mere human doctors analyze them, you know; if we so much as take a type we get fired. They all get sent aboard those damn ships, and if there's anything that needs treatment we get a report—except that actually it's more like a command from on high. They won't tell us anything about the samples, not so much as the Rh factor; that might invade somebody's privacy."

"So what?"

"I don't *know*, damn it! I just don't trust them. They're up to something; they must be. They've invaded Earth, conquered the whole damn world, and they've done nothing with it except collect blood samples and give lectures. They must be doing *something*, with these damn lectures and samples! It is the only thing, absolutely the *only* thing, that they've insisted on. It's not even just a matter of giving blood or getting out; if you won't go peacefully, they'll force you. They won't force you to eat, or bathe, or do anything, God forbid they might hurt someone or infringe his rights, but they'll get their damn blood samples one way or another, even if they have to come after you with a whole squad of their goons."

Before Starkman could say anything further the level of sound from the spaceport sank momentarily and he heard approaching footsteps. He rose.

"What is it?" Dr. Curtis asked. He didn't answer.

A moment later a young man, scarcely out of his teens, emerged from the darkness into the pale glow of the remaining lights, walking casually toward them across the bands of paving. Starkman shrank back into the darkness, away from the open door; he was certain, after hearing what Curtis had just told him, that this was an Imperial zombie sent to fetch him, though the youth wore jeans and a gray scoop-necked shirt rather than the usual blue uniform.

The doctor rose, peered past him, and saw the approaching figure. "Hello, Jan!" she called.

Starkman relaxed. "He's not a zombie?"

"No, he's one of us," she replied.

As the youth drew near he called, "Hello, Janet! Who's this guy?" He had a European accent Starkman couldn't quite place.

"Jan, this is John Starkman. Mr. Starkman, this is Jan Mueller."

Starkman nodded politely. Mueller stuck out a hand, and he took it briefly.

"Pleased to meet you," Mueller said.

There was a moment of awkward silence as the three stood in the unlit passage.

"Is Ramón here?" Mueller asked at last. "I was supposed to meet him, but I got held up at work."

"I don't know; we haven't been inside yet," Curtis replied.

"Oh." He looked back and forth uncertainly.

"You go on in," she told him. "We're still talking."

He shrugged. "Okay." He walked past and pushed through the door at the inner end, letting a wash of dim light into the passage for a brief moment.

Starkman watched him go, then turned back to Curtis and said, "All right, so the aliens really are invaders from outer space, and they've taken over the world and are taking blood samples from everyone for some unknown purpose, and you guys are a secret group of revolutionaries who want to drive them away. Right?"

"Right."

"So why did you bring me here? Why didn't you just pass me through, take a sample of my blood and give me a number like everybody else?"

She hesitated.

"Oh, come on, why me? It was something about my eyes, wasn't it? Have they got a special grudge against freaks? Or did you think that anyone as weird as I am must just naturally be a rebel?"

She flushed slightly, visible even in the dimness. "You don't have to get sarcastic."

"Why not? Why did you pick me?"

"Look, I'm a doctor, right? They pay me to examine incoming people, and to take their blood samples for them."

"So?"

"So I know a bit more about it than most people. I know that they've given us special orders to keep a sharp eye out for anyone with unusual physical characteristics, particularly peculiarities involving the eyes; and they've ordered us to keep our mouths shut about it, too. They got very excited the first time I sent them an albino. We're supposed to tag samples 'special' from anyone out of the ordinary; they get rush priority treatment, and there's a lot of fuss whenever we send one up. They also have a thing about yellow in particular; I wasn't there, but I understand the first few cases of yellow jaundice they found had them tripping over each other with excitement, sending doctors every which way and ordering tissue samples and biopsies and whole liters of blood instead of just a few cc's."

Starkman felt a sudden cold tightness in his gut.

"For years I couldn't figure out what they were doing. Then one day you come in, and I see those eyes of yours, and I'll tell you, Mr. John Starkman, there was no doubt in my mind that whatever they've been looking for, you're it."

Chapter Six

It had once been a waiting room, but a large, rough table and a metal desk had been added, converting it into a conference room and impromptu headquarters. A motley assortment of couches and chairs surrounded the table and lined the walls; most of them were empty. A television set occupied one corner, a slightly blurry newsman reciting something inaudible from its screen and a battered computer keyboard sitting atop it. There were no windows, but half of the dozen ceiling lights still worked.

There were seven people in the room, counting himself. He had already met Dr. Janet Curtis and Jan Mueller, and after he had allowed himself to be brought in, Starkman was introduced to the other four.

Emilio Suares was the group's nominal leader, a man in his late thirties and a fanatic Brazilian patriot; when Curtis and Starkman entered he was arguing with Mueller about whether Anglos would be permitted to stay in Brazil once the aliens were driven away—the extraterrestrial aliens, at any rate—and order restored. The argument was conducted in English, since Mueller spoke no Portuguese and Suares spoke no Dutch.

Raoul Santangelo was born in New York, but his native language was Spanish, and his English was heavily accented, even more so than Suares'. He was in his early twenties, and had a nasty grin that discomfited Starkman almost as much as the automatic rifle he had slung on his back.

Cheryl Fanshaw was British, and was in Brazil and in this room only because Santangelo was there, it seemed. She spoke as little as possible, preferring to cling silently to her man's arm.

Paulo Ingrao was in his fifties and sat quietly in a corner. He said nothing at all, but merely nodded in acknowledgment when his name was mentioned.

When the introductions had been made, Suares demanded, "Why do you bring this man?"

Curtis answered, "I'll let you see for yourselves. Mr. Starkman, would you take off your glasses?"

"I would prefer not to," he replied.

"But..." Dr. Curtis was momentarily confounded, her effect spoiled. She shrugged. "Suit yourself, then. Anyway, Emilio, he's got yellow eyes, like nothing I've ever seen before. It's got to be him that they're looking for."

Suares looked Starkman over carefully, then demanded, "Let me see."

Starkman noticed Santangelo's hand reaching back for his rifle, and decided that he was in no position to argue. He took off his sunglasses.

Suares studied his eyes carefully, standing him under one of the functioning ceiling lights.

"They look real," he admitted.

"They are real," Starkman said.

"In that case, Mr. Starkman, as the doctor has said, you are the man that the government has been looking for so energetically. Why?"

He shrugged. "I haven't any idea," he said.

"Are you sure?"

"Yes, I'm quite sure."

"You have had no previous dealings with these aliens? None at all?"

"No, nothing; I didn't know they existed until their ship picked me up a few days ago."

"Maybe it's just a coincidence," Mueller suggested.

Suares dismissed that possibility with a disdainful wave of his hand. "It can be no coincidence. The government searches for a man with yellow eyes, over and over and with great effort, as Dr. Curtis has told us. Now we find a man with yellow eyes. Can there be two such men?"

"If there's one, there can be two," Santangelo answered.

Suares considered that for a moment, then asked, "Are there others? Mr. Starkman, have you ever seen another with yellow eyes? Anyone in your family, perhaps?"

"No."

"Have you ever heard of any?"

He remembered all the doctors he had seen over the years of his childhood, before he got fed up and refused to see any more. They had been excited, or frightened, or merely startled. All had called him unique, or rare; none had ever mentioned anyone else with his particular peculiarity. At least one had said that he'd never heard of such a thing. "No," he replied.

"Have you heard of others, Raoul?"

Santangelo shrugged.

"Dr. Curtis?"

"No, never. I didn't know it was possible; I thought the aliens were crazy if that was what they were looking for."

"You see? It can be no coincidence. I thank God that we found you."

"Why?" Starkman demanded. "You don't know why they wanted me; why should you be so glad that you've found me before they did?"

"Because, Mr. Starkman, if you are here, with us, you cannot be there, with them. While we have you, they do not, and whatever they seek is beyond their grasp."

"How do you know that's what you want?"

"Mr. Starkman, we are sworn to oppose the alien conquerors. Whatever hampers them aids us."

"You can't be sure of that. What if they want me to help them with one of their beneficial projects?"

"In that case, we must keep you from them to weaken their support among the foolish people who are impressed by their projects. We want their projects to fail, however great the superficial benefits of those projects may be. To accept gifts from the conquerors only drives humankind deeper into slavery."

Starkman sat down on a convenient couch to think

this over, and from old habit put his sunglasses back on; he felt naked without them when in the presence of other people.

"You seem unconvinced," Suares continued. "Mr. Starkman, we are at war with alien invaders. In war, many people are hurt. We cannot accept favors from the enemy without weakening ourselves. We must keep from the enemy anything that may be of use in the struggle to defeat us in spirit as we have already been defeated physically."

"And you think I might be of use to them somehow?"

"Yes, exactly."

"I don't see how. I've been examined by a dozen doctors, ever since I was born, and there's nothing wrong with me, nothing special at all, except for the color of my eyes. My vision's been tested in every way possible, and it's completely ordinary. I was even tested for ESP, and scored just about dead average. I can't imagine what use the aliens could have for me." If, he added to himself, they really were aliens. This group seemed to accept their extraterrestrial origin, but he was still not wholly convinced. It was a topic that might bear later discussion.

"We do not know any more than you do," Suares replied, "but perhaps we may find out. Until then, I ask that you stay here with us, where we may keep you safe from the government's spies."

Starkman turned to face Dr. Curtis. "You're sure that they were looking for people with yellow eyes?"

"I'm sure," she replied. "We were told to watch for oddities, and the example they used, every time, was yellow eyes. And they did make a fuss about jaundice cases and albinos. What else could it be?"

"Were they looking for anything else?"

"Not that I know of," she answered.

"We know little about the aliens' goals here on Earth," Suares said. "They are seeking a man with yellow eyes, and taking blood samples from everyone alive. They build us cities and plant wonderful farms. Everything they do seems to be for these purposes and nothing else. The cities and farms are surely to keep their slaves happy and complete their conquest, but why do they

take the blood? Why do they seek a man with yellow
eyes? We have heard rumors that we cannot confirm
that they have done other mysterious things as well.
There is a story that they have taken hundreds of ar-
cheologists to Europe and Africa to seek for something
there. We do not know if that is the truth, or why they
should do so. We do not know why they take blood, or
why they seek you. We want to know, very much."

"Why? What are *you* trying to do?"

"I have told you, we seek to drive the invaders from
our world."

"Then what'll you do?"

Suares paused. "We have not decided that. It is too
early to say."

Mueller snorted. "He wants to send everyone home,
and let us northerners freeze while his Brazil rules the
world."

"I do not seek world rule; it is you who wishes that,"
Suares retorted.

"That's true enough. The aliens did one good thing
in uniting everybody; why mess it up? I say we should
maintain a world state, where all nationalities share
equally."

"You think such a state could survive?"

"Of course it could!"

The argument Starkman's arrival had interrupted
flared up again. The other revolutionaries joined in,
either taking sides or trying to end the debate, and
Starkman took the opportunity to wander over toward
the television. He hadn't seen one work in years—unless the pictures that had accompanied the orientation
lecture had been video of some sort.

This set was familiar enough; either there hadn't
been any major design changes, or it was an old one.
The same could be said for the keyboard atop it, though
he didn't recognize the particular make.

There was no sign of an antenna; the picture was
coming in on cable. The channel selector had one
hundred and forty-four choices, using a twelve-place
dial and twelve buttons.

He knew nothing about what was available, and the
news report would probably be as interesting as any-

thing else he was likely to stumble across; he found the volume control and turned up the sound slightly.

A babble of Portuguese came from the speaker as the announcer pointed at a red dot on a map Starkman did not recognize. He turned the volume back down and watched; he knew no Portuguese at all, and could not make out a single word.

The image switched to two men sitting in one-legged plastic chairs on a red-carpeted dais, speaking to each other. It appeared that a talk show had replaced the news.

Disgusted, Starkman turned away, leaving the set on, just as Dr. Curtis managed to put an end to the argument. Now that order was restored, Starkman decided that it was time he reasserted himself. He had been allowing others to drag him wherever they pleased—from Pennsylvania to Brazil, of all places, and from the welcome center to this rundown conference room and the company of a handful of would-be revolutionaries. Throughout, he had been treated as an object, something to be possessed, rather than as an independent human being, and he resented it. He intended to stop being treated as less than he was. He was accustomed to being a freak, but not a possession to be stolen back and forth.

"I have a few questions," he announced.

Suares, after a final glare at Mueller, said politely, "We will do our best to answer them. We wish to have your willing cooperation."

"Good." He glanced at the rifle on Santangelo's back. "I hope that you can give me a reason to cooperate." He waved toward the empty chairs on every side. "Sit down," he said.

Everyone except Santangelo sat; the New Yorker leaned against a wall instead, kicking aside a chair that stood in his way.

"You people are an underground resistance movement, right?"

Curtis and Suares both nodded agreement.

"What do you call yourselves?"

"Ah." Suares glanced at Mueller and Santangelo. "That has been a matter of debate, and we have all agreed that names are unimportant so long as we are unified in purpose. We have decided, provisionally, to call ourselves simply the Underground, a name that leaves open our methods and goals."

"That's what Suares says," Santangelo growled. "We're People's Liberation Front, man."

Curtis was about to protest, but Starkman cut her off. "You can argue later. Dr. Curtis, how did you manage to be there doing examinations when I came along?"

She shrugged. "It was just luck. We've got about half a dozen of our people among the doctors working at the Center."

"You do?"

"Sure."

"That's a lot better than I would have expected. Just how big is your organization? I thought maybe it was just the six of you here."

"Hell, no," Curtis replied. "There are hundreds of us."

"Hey!" Santangelo interrupted, "Don't tell this guy too much! He might be some kinda spy!"

"Even if he is not," Suares agreed, "he may later be captured and interrogated."

"Okay, okay," Starkman said, "I won't ask about that. Still, six doctors out of—what is it?—thirty or forty, I'd say; that's pretty good. If you're doing that well with the whole population you could start your revolution tomorrow."

Curtis glanced at Suares, who shrugged.

"We're not doing *that* well," she said. "We've been making a special effort to infiltrate the Welcome Center, and the rest of the government as well, so as we'd be able to keep an eye on what was going on. We've done pretty well there."

"They don't do any kind of security check?"

"No. I don't think they know we exist; we've been keeping a low profile so far. And all they care about is whether they get their blood samples, and they've got that all monitored."

"By computer?"

"Yeah, by Max."

"Who's Max? I heard you talking to him back at the Center."

"Max is the general-purpose computer network; he's hooked into just about everything everywhere these days. He runs the phones and the video system and handles all the data processing for everybody, including the government. He does the monitoring and keeps the medical records."

"Why didn't he protest when you told him you wanted privacy?"

She shrugged. "I don't know. We can have privacy when we want; it's a rule. They told us that right at the start, that they didn't want to infringe on our rights." She snorted. "As if we have any rights any more!"

Starkman made no comment about that, though it seemed obvious to him that whatever the aliens' true purpose might be, they were being as benevolent as they could. These people certainly still had every appearance of freedom and human rights.

"So tell me about the aliens," he said. "Where did they come from? What do they look like?"

"You've seen them," Curtis said.

"You mean the zom—the young men in uniform? Those are the aliens?"

"Those are all anyone's ever seen."

"Where did they come from?"

She shrugged again. "Who knows? Somewhere in Sagittarius, apparently."

"Are you sure they're really aliens? They look so human!"

"They came off the ships, and the ships came from outer space; there's no doubt about that. I know it's strange that they look so much like us. Maybe they changed their appearance somehow, or maybe all those old theories about gods from other worlds are right."

Starkman still didn't believe that, but it didn't seem he was going to learn anything new about the aliens' authenticity here. He changed the subject. "You don't know why they'd want me?"

"No."

"You don't have any idea?"

"None at all."

"When they've taken an interest in people before—yellow-jaundice victims, or whoever—what did they do to them?"

"Nothing very special, really; they just took one sample after another, and gave them a super-thorough exam. If it was jaundice, they cured it; if it was albinism or whatever, they just turned them loose."

"They weren't hurt?"

"No, not really. But that doesn't mean anything, because they weren't what was really wanted, and you are."

"I'd figured that out for myself, thank you." He considered for a moment, then asked, "Are those aliens human?"

Curtis hesitated. "I don't know," she said at last. "They all *look* human."

"I know. You said that maybe they'd changed their appearance; have you ever examined one?"

"Oh, no! They won't allow it."

"Do they all speak English?"

"No; some speak Portuguese, or Spanish, or whatever."

"They don't all speak everything?"

"No. Or at least, they don't admit it; I've never heard any of them speak more than one language."

"What about their native language? What's that sound like?"

"I've never heard it; I don't think anyone has."

"The ones who captured me said their names were Mike and Jason and so on; where'd they get such ordinary names?"

"We don't know, but they all have them. They all have completely ordinary first names appropriate to whatever language they happen to be speaking, and they answer to them as if they were their own real names. None of them admit to last names."

"And no one's ever seen any *alien* aliens?"

"No."

"What about the Governor I've heard mentioned?"

"He's their head man, whatever he is, but no one's ever seen him. He won't talk to anyone except his own Imperials face to face; when he talks to ordinary Earth people it's always over audio circuits, with no video. I've heard some big shots get to speak with him through one-way glass, but nobody human has ever seen him."

"Do you think he's not humanoid?"

"Who knows?"

Starkman considered for a moment. "I think they must be androids," he said. "It doesn't make sense for aliens to look and act like humans."

Curtis shrugged. "Maybe they are. We just don't know. They're not robots; they breathe and eat, and we've seen 'em bleed. And whatever they are, they may act like humans, but they don't think like humans."

The conversation was interrupted by a sudden clatter from the hallway outside the conference room; someone was running. Starkman turned, startled, as the door of the room burst open and a young man leaned in.

"Emilio! Imperiales!"

"Que?" Suares was on his feet, and Santangelo was halfway to the door by the time Dr. Curtis managed to stammer out, "What? Where?"

"There are *Imperiales,* Doctor Curtis, coming up the path from the Welcome Center. They have machines, lights, and trackers."

Santangelo was out the door, pushing past the newcomer with his rifle in his hand, then breaking into a run.

Starkman sat, utterly confused, as Suares followed Santangelo out, and Mueller vanished through another door. He had no idea what they were planning to do. Curtis and Fanshaw were on their feet but not moving, and looking as bewildered as Starkman felt. Ingrao simply sat, unperturbed, watching the others curiously.

A sudden chatter of gunfire sounded from somewhere not far away, and Starkman knew where Santangelo, at least, had been going. He wondered who had fired first.

A flicker caught his eye, and he turned to look at the television. The talk show had vanished, to be replaced by a red-and-black BULLETIN symbol. A moment later that faded away, and with a start Starkman realized that he was seeing the outside of the building he was in. Spotlights washed the plywood-and-metal walls in uneven light, and tiny figures moved.

The others remaining in the room had not yet noticed it; he decided against pointing it out.

The picture was apparently being shot from a helicopter, or perhaps a crane of some sort; the image was quite steady, with none of the vibration he would have expected in a helicopter-mounted camera. It looked down at the façade of the processing center from about twenty meters up. Starkman watched as it zoomed in slightly on two figures crouched in the doorway he had entered through; he could not distinguish faces, but judging by what he could see of their clothing he took them to be Suares and Santangelo. One of them held a weapon; he aimed, and fired at something off-screen.

Simultaneously, Starkman heard another burst of gunfire nearby. The sound was still turned off; the shots were real.

The camera zoomed in closer, and Starkman was almost certain he recognized the gunman as Santangelo. Then the image shrank again, and the camera panned outward across the surrounding pavement to the forest. Indistinct shapes were moving about in the underbrush, between the great spotlights that were spaced out along the edge of the pavement, at least four of them.

One man—or zombie, or Imperial, or alien, or whatever he was—lay unmoving in a dark pool of something, half hidden in shadow, half lit by the beam from one of the lights. Starkman felt a slight twinge of nausea. Santangelo was obviously not just playing games.

He was distracted by Mueller's return. The man's arms were heaped with guns, and he began handing them out, like a wartime parody of Santa Claus distributing gifts.

Fanshaw refused to accept one. Curtis hesitated, but finally took a small submachine gun. Starkman, who knew almost nothing about firearms, thought he recognized it as an Uzi. Ingrao silently received an automatic rifle. Mueller looked at Starkman uncertainly, then turned and left, still holding three or four guns, without offering Starkman a weapon.

Starkman honestly could not have said whether or not he would have accepted had one been offered.

Moving with serene calm, Ingrao rose and walked out the door, his weapon ready in his hands. Curtis followed him, far less composed, leaving Starkman alone with Fanshaw.

He heard the sharp crack of a single rifle shot a few seconds later, then a stuttering burst from one of the automatic weapons, then a moment of silence. He turned back to the television, noting in passing that Fanshaw jumped at the sound of every shot.

She noticed where he was looking, and her eyes turned as well. "Oh, my God," she said when she recognized the scene.

The two figures in the doorway were joined by others, the whole group jammed into the opening, their weapons bristling.

"Do you think the Imperials brought the TV people themselves?" Starkman asked.

"No, I don't think so; one of the news networks just follows them around whenever it sees a group of them. They've done it before."

Starkman nodded. Ambulance-chasing, that's all it was; there was nothing mysterious about that.

The camera panned out across the pavement and the forest. Starkman saw that half a dozen Imperials lay dead or wounded, where only one had been before. He realized that there might be others lost in the darkness or behind the trees. There was no sign that they had been armed, or that any of the gunfire came from anywhere other than the group in the doorway.

"Oh, my God," Fanshaw said again. She reached over and turned up the sound.

The newscaster's voice told Starkman nothing at all; like everyone else he had heard on this channel, she spoke Portuguese. Fanshaw listened intently, and Starkman wondered whether she knew the language or was just trying to pick out familiar words where she could.

The momentary silence outside was broken by a calm voice, distorted by amplification and intervening walls, eerily echoed by the television, speaking English loud enough to be heard in the windowless room without any relay.

"I am speaking as a representative of the government. We mean you no harm. We respect the right of all citizens to assemble, and the right to bear arms. We have come for the new arrival taken from the Welcome Center, the man who called himself John Starkman."

Gunfire interrupted the speech. Someone was shouting, incoherent through the corridors and doors, and not loudly enough to be picked up by the television crew. The newscaster was saying something in Portuguese. Starkman sat, frozen, trying to accept the fact that the voice had said what it had. He could handle the idea that they had come after him, but the statement of rights contrasted so grotesquely with the twisted bodies of the downed Imperials that the camera kept returning to that he could not believe both speech and bodies were real and simultaneous.

Starkman decided that it was time to leave. Dr. Curtis had been quite correct; the government did want him for something, and he didn't want any part of it until he had a better idea of what was happening. Furthermore, he didn't think he wanted anything to do with either side of this conflict, neither the revolutionaries blazing away at anything that moved, nor the government that blithely sent unarmed men up against guns while reasserting its commitment to human rights.

The revolutionaries hadn't thought to leave anyone to guard him, unless the young Englishwoman, unarmed and obviously frightened, was supposed to do that. Her eyes were fixed on the set, and paying no attention to him.

Of course, he couldn't go out the way he had come in; that doorway was where the gunfire was coming from, where the firefight was blocking his way. He was sure that a building this size must have other exits; it was just a matter of finding them and eluding whoever might be guarding them.

As he thought that over, the camera panned away from bullets bouncing harmlessly off the armor plastic of the spotlights, and zoomed back for an overall bird's-eye view of the processing center. Starkman saw that the Imperials and their machines were all clustered together in a ragged pool of light at the front of the building, facing the door where the Underground waited, while the rest of the area remained dark.

That seemed incredibly stupid; in fact, both sides were being stupid. There *must* be other doors; why was there no sign of anyone guarding them?

Perhaps they were sealed off, somehow, or the government forces there were hiding. It might be impossible to avoid the Imperials, whoever and whatever they really were. However, if he stayed where he was, he would almost certainly be either captured or killed; the revolutionaries couldn't stand off the government forever, he was sure. Capture might be tolerable, but he had no intention of getting killed if he could possibly do otherwise. If he left the building by another route, he might be captured or he might escape, but he didn't see any reason for the Imperials to kill him.

Having thought this out, he stood, hesitated, and then decided to try the door he had entered by, and which most of the others had departed by. He crossed the room and leaned cautiously out through the doorway.

He waved at Fanshaw vaguely, hoping she wouldn't press the issue; he would have had to shout to be heard over the gunfire in any case.

The corridor was empty; to his right were the swinging doors that separated it from the passageway where he had sat and talked with Dr. Curtis, and where the fighting was now going on. To his left the corridor ran off into the dark for about twenty meters and then ended in a blank wall. There were two doors in the opposite wall, and two others, besides the one he stood in, in the near wall. All were closed.

He wished that the old waiting room had windows, so that he could have seen directly what was going on outside. The swinging doors *did* have small windows in them, through which light was pouring into the corridor from one of the government spotlights. He crept up to peer through, his sunglasses protecting his eyes from the glare.

He was looking at the backs of several of the revolutionaries; he could see Santangelo and Curtis and Ingrao, and a figure he didn't recognize immediately, then realized was the young man who had alerted them to the approach of the enemy. Suares was presumably in the front of the group, hidden from him by the others. They were all jammed together in the passageway, taking turns in leaning out through the entrance and firing, it seemed. He could not see past them to see what they were shooting at; for all he knew they were firing at nothing.

This was not a way out. He turned and headed up the corridor.

The doors opposite the waiting room opened into a lecture hall, much like the one he had sat in at the Welcome Center, though somewhat larger. It seemed to him that the similarities might continue; perhaps there would be exits at the front of the hall leading to examining rooms, and some way of leaving the building beyond those. He took off his sunglasses, which were a serious impediment in the dark, and started down one of the sloping aisles.

He heard footsteps behind him, suddenly audible in a lull in the gunfire; he turned and found Mueller pointing a rifle at him.

"Where did you think you were going?" Mueller asked.

Starkman shrugged.

"Get back out here."

Reluctantly, Starkman obeyed, emerging once again into the corridor. He fumbled with his glasses, uncomfortable with them off, but not sure that there was enough light to see by with them on.

Mueller began to say something, but was interrupted by the amplified voice from outside.

"We regret that you have chosen to behave violently, rather than turning John Starkman over to us peacefully. To insure his safety we must take him from you by force if you do not surrender him to us immediately. We have summoned human troops to aid us, since we are categorically forbidden to harm any human beings ourselves, whatever the circumstances."

There was a pause; Starkman heard Dr. Curtis shout, "Traitors!"

The outside voice spoke again. "We offer you a last chance for a peaceful settlement. This is your final opportunity to avoid bloodshed. We will not harm any of you, nor will we harm Mr. Starkman, if you surrender now. We can make no guarantees of your safety if you continue to resist. You have fifteen seconds to reconsider."

Curtis shouted a string of obscenities, and someone fired a burst from one of the automatic weapons.

Then there was a moment of silence. Starkman, staggered by the voice's offer of an "opportunity to avoid bloodshed" and the memory of the bodies he had seen on TV, mentally reviewed once again what he had seen and heard. All the gunfire had been from the members of the Underground; there was still no sign that the government's forces, whoever they were, had returned fire. This government seemed to have a perverse idea of what constituted bloodshed; it apparently counted only its enemies.

He wondered if the revolutionaries had noticed that; he was about to ask Mueller when there came a sudden whistling noise, followed by a dull thud, then renewed shooting and shouting, all in the little entryway. Mueller turned, gun ready, and Starkman's gaze followed his.

Something gray and cloudy was blocking the view through the little windows, and what looked like pale smoke was seeping under the swinging doors.

Another whistle and thud sounded. "Gas!" Mueller spat. He lowered his gun and ran toward the door.

Starkman immediately ran in the opposite direction. As he turned and dove through the farther of the two doors to the auditorium he saw Mueller push open one of the doors, releasing a billowing cloud into the corridor. The shouting beyond had turned to coughing, but the revolutionaries were still firing sporadically.

Another whistle sounded, much louder this time; Mueller was holding the door open. The thud came from the far end of the corridor, and a new cloud sprang up.

Starkman ignored it and ran down the aisle, slam-

ming into the front wall before he could stop himself
completely. There was a door to one side; he struggled
for a moment with the latch, then got it open and staggered through. A thin fog was drifting through the lecture hall as he did so; he felt a tickling in his throat.

He was in another corridor, dimly lit by a single
surviving ceiling light and lined with doors on both
sides. He picked one at random.

It was locked.

He tried the next, and the next; they were all securely locked, and would not yield to his rattling and
pounding.

The wafting smoke was beginning to seep into the
corridor; from somewhere behind him he heard muffled
shouting. The sound of gunfire had ceased completely.

Desperately, Starkman flung himself against the last
door in the corridor. The shoddy construction of the
building worked to aid him; it sprang open, and he
sprawled heavily into the dark room beyond, while a
broken piece of door frame clattered off the far wall.

He picked himself up off the floor, found that his
sunglasses were still in his hand and miraculously unbroken, then paused and looked around.

He was, as he had expected, in an examining room.
A folding screen leaned against one wall, and electronic
equipment was piled haphazardly to the other side, apparently discarded. The machinery was perhaps a third
of what he had seen in the room where he had met Dr.
Curtis; the remainder, he guessed, had been worth sal-
vaging, and had been removed when the building was
abandoned.

Noticing that the room was unlit, it occurred to him
to wonder why the ceiling light in the corridor was on.
Had the revolutionaries turned it on, and then left it?
Did it come on automatically when someone entered?
Had it been left on by accident?

It didn't matter, save that it might possibly have
implied that someone was in this part of the building,
and he thrust the thought aside.

By the feeble light from the corridor, he saw that
there was a door in the far wall, as he had expected. It
was locked, but with a simple bolt on his own side of

the frame. He drew the bolt, turned the knob, and stepped through into another corridor. Like the examination room, it had no artificial illumination, but a faint trickle of pale moonlight worked its way through a dirty window at one end. He ran toward the light, coughing.

The window was in a door, and the door opened readily to his push on the release bar.

He was outside, on the side of the building directly opposite where he had entered, the far side from where the fighting had taken place. From where he now stood there was no sign of violence, and the only trace of gas was a thin wisp that drifted out the door behind him before he pushed it shut.

He was on a narrow strip of pavement; beyond it was more of the forest, and faintly visible through the trees he could see the lights of a city in the distance. Off to his left were the other two buildings of the complex. He was in the shadow of the building he had just left, with the moonlight and the peripheral glow of the spotlights spilling down past onto the forest, leaving him shrouded in blackness. There was no sign of Imperials or any other forces, government or otherwise, so far as he could see.

A shadow moved, and he glimpsed something at the edge of his vision. He looked up.

Floating overhead, gleaming in the moonlight, was a machine, perhaps two meters wide and five meters long. Startled, he stepped back toward the building.

A man's head appeared, leaning over the side of the hovering machine. "Hello," a friendly voice said. "Need a lift?"

Chapter Seven

Starkman stared upward in openmouthed astonishment.

The man smiled at him, then drew back out of sight; an instant later the machine began settling lower. As it descended, the man's head reappeared, and he said, "I think we'd better hurry. Right now all the cops are around the other side, because that's where all the shooting is and I told the people back here to go help, but any minute now one of them is going to remember that there are other entrances and they're supposed to have the place surrounded. Either that, or someone is going to realize that I'm not supposed to be here and haven't got the authority to send them anywhere. They may be stupid, but they'll figure out where you've gone pretty quickly when they see you're not in the building."

"What are you talking about?" Starkman was thoroughly confused.

The machine was down to eye level now, and he was able to look it over. It was obviously a vehicle of some kind; its upper surface strongly resembled an old-fashioned convertible, with brown upholstered seats for four surrounded by curving metal that gleamed red even in the moonlight. There was no sign of a camera, so he dismissed the thought that it might be connected with the television crew. Besides, the man spoke American-accented English, not Portuguese.

"I'm talking about getting you away from here before those idiots the government sent after you find you. Anywhere you can go on foot they can track you, but

I'm pretty sure I can lose them with this baby." He patted the smooth side of the machine.

"But who..."

"Look, we can talk later, but they'll be here any second; jump in!"

A crash sounded somewhere in the building behind him, and Starkman made his decision. He leapt over the low side of the vehicle and landed in moderate disarray in one of the rear seats, almost losing his sunglasses. Before he could untangle himself and put his feet on the floorboards the machine's pilot did something with his right hand and his feet, and the thing took off forward, rising, at an acceleration that effectively pinned Starkman where he was, knees bent, right arm draped over the rear deck, left hand clutching his glasses, feet on the edge of the seat beside him. His hair whipped around in the wind, slashing across his face for a few seconds before settling into a rippling stream to the rear.

Fighting the acceleration, he got his feet in front of him and his hands on either side, gripping the edge of the seat, just in time to be flung off balance and topple heavily against the side of the vehicle as it snapped through a right-angle turn in midair.

When he had recovered from that he managed to locate a seat belt; his first three attempts to buckle it were interrupted by new and sudden changes in course, but the fourth try ended with a satisfying click.

Feeling somewhat more secure, he looked over the side and saw moonlit rooftops flashing by a hundred meters below. The landing field's lights were still visible on the horizon to the rear, glowing faintly through the fringe of forest. He glimpsed the brighter, uneven lights that played on the abandoned processing center, and thought he could make out something hovering above it that might have been the television crew, though he could not be certain. There was no sign of the Welcome Center; he guessed it was unlit.

A glittering tower flashed by, lights ablaze, and he flinched involuntarily at its nearness. The craft had passed its upper floors at a distance of four or five meters, but at this speed and altitude, at night, that seemed far too close for his liking.

There were other illuminated towers ahead and to the sides, and a carpet of lights below; they were obviously entering a city, and Starkman realized as he saw the lights approaching that they had dropped considerably lower without his having noticed it. They were no more than forty meters up. He had been distracted by trying to see where he was and where he had been.

The pilot sent the vehicle around another sharp turn, and the whistling of the wind decreased somewhat; they were slowing.

Another vehicle, very much like the one he rode in save that it was blue in the light from below, rather than red, zipped overhead.

That reminded him that he didn't know anything about his own craft or its driver; he took a few seconds to look them over.

He was sitting in a comfortable, velvet-upholstered bucket seat; another identical seat was to his right, and two more in front. The driver sat directly in front of him; the other two seats remained empty. Strapped in as he was, he was unable to see past the pilot's shoulders to study the controls; he decided, as the thing whipped around another corner, that he could wait until he was safely on the ground before investigating the controls. His stomach was churning uncomfortably as it was.

To his right, between the back seats, there was a hatchway or lid, presumably leading to a storage compartment. Under his feet the floor was ridged black plastic. The inner surface of the craft's side, to his left, was covered in dark fur, which felt artificial. A narrow band of chrome ran around the lip of the passenger compartment, sparkling gold and silver in the city lights, and beyond it was the exterior that shone bright cherry-red whenever they passed near a bright light. The vehicle, Starkman decided, would be very flashy by daylight.

The exterior deck extended a meter to the rear and perhaps a meter and a half to the front, and perhaps five centimeters to either side, before curving down out of sight. There was a gracefully curved windshield at the front of the passenger compartment; he was unsure how much use it actually was in the absence of side windows and a roof.

All he could see of the driver was a pair of broad shoulders clad in black vinyl and the back of a head covered with thick short black hair.

That seemed to complete his inspection of the vehicle. He still had no idea what powered it, how it flew without wings or rotor, or who his savior was. He looked over the side again.

They were flying over a city street and had dropped to scarcely a dozen meters. He judged their speed to be in the neighborhood of eighty kilometers per hour, and dropping. Below, street lights and neon signs flashed by, illuminating pedestrians in bright clothes, building façades, and a steady stream of traffic.

The vehicle tilted sharply under him as the pilot threw it into another curve, turning onto another street and missing the building on the inside corner by less than a meter. Starkman suppressed a gasp as the dark concrete wall swooped near, then receded.

They slowed further, maintaining the same altitude, until they were dawdling along at about thirty kph—which was still far faster than Starkman had traveled in an open vehicle in a decade or more. He was sure the ship that had brought him to Brazil had moved much faster, but he had felt none of the acceleration and seen nothing of the landscape rushing by. He felt more than a little battered and ill from this ride.

The pilot fiddled with something, then leaned back and said over his shoulder, "I've got us cruising on automatic; we can talk now."

Starkman was recovering his aplomb, more with each decrease in speed, though he was still unsettled. He shouted, "Who the hell are you?"

"Oh, don't worry about that; I'm just a friend, here to watch over you. My name wouldn't mean anything."

"Are you with the Underground?" He gestured back toward the general area where he guessed the processing center to be.

"Well, no, actually. I had originally planned to lie and tell you that I was, but that was because I thought

I'd be picking up some of them, as well as you. It was a real stroke of luck getting you alone; made it a lot easier to get you away from that bunch of hotheaded idiots."

"How'd you know I was the one you wanted?"

"You've got to be a new arrival; who else would be wearing a ratty winter coat in this weather?" He waved at the warm golden moon and blue-black sky above the buildings. "And you had to be John Starkman, because that's the only new arrival they had with them." Starkman had to admit that it was a logical deduction. The air was fresh and still warm, though it was a good two hours, at least, since the sun had set.

"Who are you?"

"I told you, don't worry about it."

"Are you with the government?"

"No!"

"Then where'd you get this ship?"

"What ship?"

Starkman pounded on the back of the driver's seat. *"This* ship! This vehicle!"

"It's not a ship, it's an aircar!"

"Whatever it is, where'd you get it, if you don't work for the government?"

"I bought it, of course."

"Oh." It had not occurred to him that such a vehicle might be available to the public at any price. He had somehow assumed that all technological advances that had been made in the last dozen years, whether of terrestrial or extraterrestrial origin, would be government monopolies.

A siren sounded, and something dark flashed overhead; the siren's howl faded as the other aircar dwindled into the distance.

"Why did you pick me up?" Starkman asked.

"To keep the government from getting you back." The pilot leaned forward, fiddled with something, and swung the aircar ninety degrees onto a new street. That done, he sat back again and added, "I don't want them to get you any more than those idiot Undergrounders do."

"Why not?"

"Well, that's complicated. I'd rather let my employer explain that."

"What employer?"

"You think I work alone? I didn't just turn up at the old processing center so fast by dumb luck, mister; I'm with a group that has its own ways of knowing what's happening."

"What, some rival underground?"

There was a pause before the pilot answered, "I'll let my employer explain that one, too."

"So you work for someone who told you to come and rescue me from the government, and told you where I'd be?"

"That's roughly it, yes."

"How do I know that this isn't some trick by the government to make me cooperate? It seems pretty damn unlikely that you were able to just fly this thing right up to the door without being stopped."

"You just don't appreciate how stupid the government's people are. I told you, I told them I'd been sent to replace their guards at the back."

"All right, maybe I don't appreciate their stupidity, but you didn't answer the question; how do I know you don't work for these supposed aliens?"

There was a long pause before the man replied, "I never said I didn't work for the aliens; I said I didn't work for the government."

"What's the difference? You mean you *do* work for the aliens?"

"I work for an alien, yeah. Did it ever occur to you that maybe they're not all alike? Maybe they aren't all working for the Governor?"

Starkman considered that for a moment, as the pilot took the aircar around another corner, narrowly missing an aircar coming around the same corner from the opposite direction.

"Then who the hell *are* they working for?"

"I'll have to let—"

"I know, I know, you'll let your damn employer explain that."

"Right."

"So when do I meet this employer, and hear all these explanations?"

"That's where we're going now, and we're almost there." He gestured vaguely ahead.

There were still several questions Starkman wanted to ask, but he was tired of shouting over the wind; even at low speed the open vehicle was not the best place for a casual chat. He was also, he decided again, tired of being dragged hither and yon by one group after another, none of them anyone he could trust. He didn't know enough about the situation to choose sides; he didn't even know yet what the sides really were. What he wanted was time to himself, time unbeset by Imperial zombies and gun-happy revolutionaries and maniac drivers, time in which he could look about himself and think things over.

At the very first opportunity, he promised himself, he would get away from this stranger with the mysterious alien employer. He would look over this city, wherever it was, and give himself time to think. Then he might come back to wherever he was being taken right now, or he might turn himself in to the government, or he might try to hunt up the Underground. Or, more probably, he might just fend for himself, as he always had, either here in the midst of civilization, or by making his way back north.

The aircar was slowing to a stop, and Starkman looked down over the side. They were sinking gradually downward toward a small white-paved parking lot, with spaces for half a dozen vehicles, and with two of those spaces occupied. The lot was enclosed on two sides by a gleaming tower of white concrete and blue glass, and on the other two sides by strips of greenery. Beyond the hedges and flowers the city streets were filled with ordinary ground cars, such as Starkman had ridden in and even driven before things fell apart, but the lot had no openings to allow such mundane vehicles entrance; it was obviously reserved for those fortunate enough to have aircars. Judging by the seething mass of earthbound metal and the rarity of the aerial equivalent flashing overhead, Starkman guessed that aircars were still an expensive luxury.

The craft that carried him settled neatly into one of the empty spaces in the lot, and he looked for a door handle.

He couldn't find any, and, following the pilot's example, he stood up and clambered over the side of the vehicle. It wasn't particularly difficult once the aircar was firmly on the ground, its silent power apparently shut off. With no wheels holding it up off the ground, he realized, the sides could easily be made low enough to render doors unnecessary.

The pilot picked a few items out of the car, stuck his keys and one or two other things in the pocket of his vinyl jacket, slung something like a camera case over his shoulder, and started walking toward a door into the building. "Come on," he called.

Starkman looked at the aircar for a second, decided he didn't know enough to steal it, and turned away. Casually, without glancing at the pilot, he walked over to the side of the lot and stepped through a gap in the hedge onto the sidewalk beyond.

"Hey!" called his escort. "Where do you think you're going?"

Starkman waved and kept walking. The traffic was too heavy to risk crossing the street, so he turned right, choosing his direction more or less at random, and strolled along the sidewalk.

"Hey, wait!" The pilot turned away from the building and dashed out to the sidewalk.

Starkman ignored him, and looked about at the other pedestrians, noticing that clothing hadn't changed as much as he might have feared. People still wore jeans or slacks for the most part, though some were cut oddly. He saw a few older women wearing skirts that reached mid-calf, a style he found utterly unappealing. The upper torso was more variously garbed; the current fashion among young women seemed to be a sort of lace-front halter, usually in bright warm colors, while men favored short-sleeved, scoop-necked shirts of flowing pastel materials. Older citizens often wore more conservative clothing, though he saw no one in anything with a real collar; his heavy plaid flannel shirt would stand out at least as much as his battered coat, though his pants would pass readily. Hair seemed to be worn short and straight; after getting himself a new shirt he'd want to have his trimmed. He did not want to stand out.

The street blazed with light, from street lamps and signs, neon or otherwise. There was obviously an extensive and active night life in this city; the street was as crowded at what he judged to be midevening as he might have expected at noon.

Or perhaps, he told himself, it just seemed that way to him. He hadn't seen a city crowd, either noon or midnight, in more than a decade.

"Hey, Starkman!" called the pilot from behind as he grabbed at Starkman's arm.

Starkman shook him off. "What do you want?" he asked, trying to sound as if he were an ordinary passerby accosted by a stranger.

"What the hell do you think you're doing?"

"What business is it of yours?"

'I brought you here so you could meet my employer, not to go running off on your own."

"Maybe that's why you brought me, buddy, but that's not why I came, and I never agreed to meet anybody. I got in your car because it looked like the only way I was going to get away from the fighting, and you didn't put any conditions on it at the time."

"Where are you going?"

"That's my business, not yours; I can take care of myself."

The man grabbed Starkman's arm again, and this time refused to be shaken off. Starkman stopped walking and turned to face him.

"Let go of me," he said, his voice low.

Recognizing the danger implicit in Starkman's tone, the pilot let go.

"I really think you should come back with me," he said.

"Look, whoever you are, I've had just about enough of being told what to do and where to go. I'm tired of it. A week ago—hell, less than that—I was living happily enough, alone in Pennsylvania, bothering nobody, doing no harm to anyone. Since then I've been kidnapped no less than three times—by those damn zombies and their talking spaceship, by a paranoid guerrilla doctor, and by you. I've been dragged from Pennsylvania to Brazil, a place that I have never wanted to visit, and hauled from spaceport to indoctrination center to revolutionary cell to here. I've had it, do you understand that? I'm not just a piece of meat to be carted around, I'm a free human being! I'll go where I please, when I please. Now, get lost!"

"No, look, Mr. Starkman, please reconsider. I don't want to have to force you—"

"And just how would you force me? I don't see anything that looks like a weapon, and I'm just about your size and probably a lot meaner than you; I don't think it's at all clear you could take me anywhere I didn't want to go. And besides that, weapon or no, look around you." He gestured at the street lights and signs and the pedestrians strolling beneath them, smiling and laughing and talking and arguing. "If you were to try and drag me and I were to put up a fight, or call for help, what do you think would happen? Do you think that all these fine people would just let you take me? Maybe they're all Brazilian and don't speak English, but a cry for help is pretty easy   figure out in any language. I don't think you're taking me anywhere I don't want to go, Mister No-Name."

There was a pause as each stared at the other, Starkman at the pilot's eyes, the pilot at his own reflection in the mirrored sunglasses.

"If you call for help," the pilot said at last, "you'll attract a lot of attention. Don't forget the government is looking for you."

"I haven't forgotten anything. I'll worry about it myself."

"I really think that you should come back with me."

"I don't care what you think."

"You're acting like a kid having a tantrum!"

"What if I am?"

The pilot was at a loss for a reply to that; he stammered for a moment, while Starkman turned on his heel and began walking again.

The other man followed, walking a pace behind and slightly to one side.

They came to the corner, Starkman in front, the pilot a pace behind, and Starkman looked up at the traffic lights. They were familiar, no different from those he had seen as a child, and he wondered how they affected aircars; did aerial vehicles just ignore them, even when following the city streets?

There were no WALK lights, nor the red-and-yellow combination that had served in Boston when he was very young; pedestrians crossed on the green, dodging whatever cars happened to turn.

Out of practice as he was in dealing with traffic, Starkman made his way across in a series of awkward dashes; when he miscalculated one run a wheezing ancient gas-burner came to a screeching, honking halt half a meter away from his foot. In a final sprint he gained the far sidewalk.

The pilot was there beside him.

"What the hell are you doing here?" Starkman demanded. "Why don't you go home?"

"I'm going to stick with you for a little while and see if maybe you'll change your mind."

"I'm not going to change my mind."

The pilot shrugged.

"Look, get lost! I don't want you around!"

"It's a free world; what're you going to do about it?"

Starkman's fists clenched, but he held himself in check. Punching the man would be satisfying, he was sure, but it would also attract a great deal of attention and possibly get him arrested. He did not want that.

Annoyed, he pried his fists open and walked stiffly onward.

They were passing a row of shops, and the warm glow from the glass front of a coffee shop reminded him that he hadn't eaten since he left the starship several hours earlier. He stopped and looked in.

There was the familiar counter and row of stools, like any diner or coffee shop anywhere in the past century, with a surly black-haired waitress in a tightly laced orange halter serving a handful of customers. Behind her, rather than the open grill and beverage dispensers Starkman was accustomed to, was a machine that looked like a cross between a computer and an Automat—he knew what an Automat was from pictures and descriptions, though they had been pretty much obsolete by the time he was born.

The door, a meter ahead of him, opened, and a man marched out, flicking an imaginary crumb from the abbreviated sleeve of his silky blouse. Starkman caught the door as it started to swing shut, but before he could step inside the pilot caught his arm and asked, "You have money?"

"What?"

"Do you have any money? This place doesn't hand out free food, you know."

Reluctantly, Starkman stepped back and let the door close itself. His step firm and resolved, he marched on.

Following up his advantage, the pilot asked, conversationally, "Do you know where you're staying? Have any friends who can put you up? Without luggage, and dressed like that, you're not going to get a hotel to take you unless you pay in advance."

"Oh, shut up," Starkman replied. The comment on his clothing struck home; he glanced down at himself, then looked around at the other pedestrians on the streets.

They roamed in pairs and small groups through the yellow pools of street lighting, entering and leaving clubs and restaurants and shops. Several cast curious or disparaging looks in his direction. He was quite obviously a new arrival; he knew it, they knew it, and he knew they knew it.

Before very long he was going to be very much an outcast, wandering the streets with no money and dressed oddly, with no place to go and no crowds to hide in when the shops and bars closed. He didn't know the language; he had heard enough of the conversation around him to be sure that most of the people spoke either Spanish or Portuguese, with only an occasional couple using English.

He didn't even know for sure the name of the city he was in. He thought that it must be the capital, which was apparently called simply "Capital," but he couldn't be certain.

There was a sudden crackle from somewhere to his right, from one of the shop fronts. He started slightly.

"The public address system," the pilot explained. "They've got them everywhere. They use them for news bulletins, calls for volunteers for new colonies, that sort of thing."

Before Starkman could reply the crackle turned into a voice, speaking in the same calm flat tone used by the starship's guiding intelligence and the speaker that had asked the Underground to surrender, though it was not exactly the same voice. Starkman assumed that that tone indicated it was a machine speaking.

It spoke in Portuguese first, then Spanish; Starkman could not follow either one, but to his horror he heard his name mentioned in both announcements. Finally, after a brief pause, it said in English, "Attention! Citizens! The government has reason to believe that a man calling himself John Starkman may be in your area. This man is a new arrival, brought here from Pennsylvania, who left the Welcome Center before he could be treated for a new and virulent form of influenza he is believed to be carrying. This mutated virus is believed to be invariably fatal within five days of infection. Early symptoms include severe respiratory congestion, nausea, headache, and cramps, followed by a dormant period, followed by high fever and spasms resulting in death. The missing man is believed to be in the dormant phase, and may think himself recovered. If he is seen, do not approach him, but report all sightings immediately to the nearest representative of the government. John Starkman, if you hear this, we ask you to turn yourself in. No harm will come to you, and your condition can be cured. The missing man is described as follows: Approximately one hundred and eighty centimeters tall, weighing seventy-five to eighty kilos, with long dark-brown hair and beard. He was last seen wearing a brown winter coat, an old-fashioned flannel shirt, denim pants, boots, and mirrored sunglasses. Anyone seeing a stranger fitting that description is asked to report the sighting immediately, and reminded not to approach him. Thank you, citizens, for your cooperation."

The message began again, in German.

Starkman looked around; pedestrians were staring at him. Worse, some took a quick look and then vanished into buildings, almost certainly calling in the sighting.

Starkman prided himself on his common sense, and he knew when he was beaten. He turned to the pilot and said, "All right, you, if you can get me out of here without being stopped, I'll meet your damned alien boss."

The pilot smiled and said, "This way."

Chapter Eight

The pilot swapped coats with him to make him a little less obvious, and then led him quickly back to the building by the aircar lot, dodging in and out of shadows and avoiding other people as much as possible. By the time they entered the building the message had gone through its German version and begun on Russian. While they waited for the elevator it was repeated in Chinese, and the first few words of Japanese were cut off by the closing of the elevator door.

Starkman found it slightly incredible that he had not already been apprehended by the police, or a squad of Imperials, or whoever enforced the law in this city. He muttered something about it under his breath just as the floor of the elevator pushed up under his feet.

He had forgotten what a ride in a high-speed elevator was like, and his stomach was still unsettled from the wild aircar ride; he clutched at the rail on the wall with one hand and at his belly with the other.

The pilot, blithely unaware of his discomfort, replied to his private comment. "I think they aren't trusting human police on this one, and there aren't any of their troops in the area yet." The car came to a smooth, silent stop at the forty-fourth floor; he turned to point the way and noticed Starkman's appearance.

"Are you all right?" he asked with sudden concern.

Starkman managed an uncertain nod; the deceleration had upset his stomach even more than the acceleration, but now that he wasn't moving he thought he would recover.

"Lord, your face is white!" the pilot said.

Starkman nodded again, and stepped out of the elevator onto thick red carpeting. He was regaining his composure. He was annoyed. A simple elevator ride, he told himself, shouldn't bother him!

With that thought in mind he managed to force himself to walk steadily and smoothly beside the pilot. By the time they reached the door marked *44F* he was back to normal. After a moment of fumbling with the key the pilot got the door open; he held it while Starkman entered.

He was in a living room that, although ordinary enough in most regards, was furnished in a style that wasn't quite anything Starkman was familiar with. There were two couches, lush red carpet on the floor, a video screen on one wall, video projector and computer console on a cabinet opposite. A row of windows ran along one wall, and a kitchen opened off the opposite end. Two doors in the far side presumably led to other rooms.

The only thing really out of the ordinary was a large wooden crate shoved into a corner. One side had been pulled open, and it was obviously empty.

Starkman was mildly surprised to find himself in an apartment; he had been expecting an office.

"Have a seat," his escort said, motioning toward the couches. He himself walked directly to the computer console.

"I thought we were here to meet your boss," Starkman said.

"We are."

"Then what are you doing with that computer?" He did not take the proffered seat, but stood in the center of the room.

The other did not reply immediately; he was working loose one of the plugs that connected the computer to the wall. It was not the power line, Starkman noted, but an input jack.

A thought occurred to him.

"You aren't working for a machine, are you?"

"What? Oh, no. My employer is no more a machine that you are. He prefers to communicate through this, though, rather than meeting face to face." He pushed a final switch and the video screen lit, a blank expanse of light blue.

The pilot ran his fingers over the keyboard, and the screen read, I BROUGHT HIM.

There was a brief pause, then IT IS GOOD appeared.

"There," the pilot told Starkman. "It's all yours. The Watcher is on the other end of the line; ask him anything you want."

Starkman glanced at the machine, then said, "I can't type."

"Don't worry about it; just peck out what you want to say."

"No. Look, mister, you told me that I was going to meet your employer, an alien. You didn't say anything about exchanging pleasantries with an ordinary computer; that thing looks about as alien as Sears and Roebuck."

"What did you expect?"

"I expected to meet an alien—at the very least one of those zombie Imperials who might know a bit more than the morons who picked me up."

A third line appeared on the screen, reading, WHY DO YOU NOT COMMUNICATE?

JUST A MINUTE, the pilot typed.

He turned back to Starkman and said, "The Watcher's a real alien, not one of their human clones. He can't breathe our air and he doesn't speak English; we communicate through the computer translator. If you want to talk to him, you'll have to talk to the computer. If typing really bothers you, I can hook up a voder."

"Hook up a voder and let me see this alien, and I'll talk."

"Why do you want to see him? How do I know you won't try to kill him?"

"Why would I want to kill him?"

"How should I know? Xenophobia, maybe, or maybe you're working for the Governor after all. I don't trust you, Starkman; you're unpredictable. You punched out one of those poor dupes on the landing field today, you skipped out on your buddies in the Underground, and you tried to skip out on me. I don't know what the hell you think you're doing, but I don't see why I should trust you."

"I suppose I should trust you, instead? You won't even tell me your goddamn *name!* How do I know that there are really *any* aliens on Earth at all? Everybody's been telling me there are, but nobody's ever seen them! The whole thing looks like a big fraud to me, a fancy hoax to keep this new world government in power. You show me a real living, breathing alien—and I don't care what he's breathing—something that's not just a computer or some weird mutant animal or some Hollywood special effects, and maybe I'll start believing enough of what you tell me to trust you! All I ever wanted was to be left alone; I didn't ask to be dragged here. You show me your damn alien or I'm leaving; maybe the government will make good on their promise to ship me home when they're done sampling my blood."

He turned and started toward the door. "Wait a minute," the other man called. "Maybe you're right. Hang on a minute." He punched buttons.

The screen read, HE WANTS TO SEE YOU.

There was a brief delay before the answer appeared, reading simply, WHY?

HE DOES NOT BELIEVE YOU EXIST. HE ACCUSES ME OF USING A COMPUTER TO DECEIVE HIM.

Again there was a delay.

THE REQUEST IS ACCEPTABLE. PERMIT HIM TO ENTER.

The pilot read the answer and shrugged. "He says it's okay. Hold on." He pulled his keys from his pocket— he had transferred them to his pants when he traded jackets—and unlocked one of the doors. He stood back and gestured. "Go on in," he said, "But don't try to touch him. I'm going to hook up the voder for you. It's straight ahead."

Starkman nodded an acknowledgement, then opened the door. Beyond it lay a short passageway with a door to one side that opened on a small bathroom, and a closed door directly ahead. He stepped in and opened the other door. Behind him the pilot called, "Remember, I'm right here and that's the only way out; if you hurt him I'll kill you."

The second door swung open and revealed a small

bare room. The walls were blank white, and the only furniture he saw was an object that bore a vague resemblance to a computer console. It had an array of colored things that might have been buttons, and a holographic sphere floated in the air above it, flickering dull red.

Sitting—or perhaps standing, or crouching—near the panel of buttons was the alien. It waved a tentacle in greeting.

"Hello," Starkman said, his throat suddenly dry.

It whistled something in reply.

He had not really expected the alien to be there, had not believed it really existed, and had not given much thought to what it might look like. Despite his own intellectual disbelief in parallel evolution, he had unconsciously assumed that it would be humanoid. It wasn't. He might have guessed that it would resemble a bird, or a jellyfish, or an octopus. It didn't. It looked, quite reasonably, like nothing Earth had ever produced. It was so utterly strange that it took him a good fifteen seconds to realize that it had some similarities to earthly creatures, to a turtle, perhaps, or to a snail.

Its body was greenish brown, or perhaps greenish gray, where visible, but most of it was covered with overlapping plates of reddish armor arranged in three sheets, one on top and one on each flank, giving it a boxy appearance. This armor was its only resemblance to a turtle. It was watching him through four stalked green eyes, each about the size and shape of a somewhat flattened baseball, on stalks the size of a woman's forearm that protruded from the joints where the armor plates met at one end. That end, he decided, was the front; in addition to the eyes there were several short slender tentacles, each ending in a jointed fingerlike structure, and there was also a lumpish something that he thought might be its head. One of the stubby tentacles was fumbling with something out of sight; others were poised over the computer keyboard. There was an orifice of some sort, with elaborately-irising lips, on the top of the head—if it was a head. Below the tentacles a thick limb something like an elephant's trunk, save that it was green and unwrinkled, showed. Below that

were two things that it rested on, which looked to Starkman rather like a snail's foot, except that there were
two. It was those feet and the stalked eyes that reminded him of a snail.

At the other end of the armor plate were two more
vaguely tentacular things that flared at the end; they
were not fully visible, and Starkman had no idea what
purpose they might serve. They were more nearly gray
than the rest.

It had no hair or fur or feathers or scales, and the
plates looked more like sandstone than like shell or
bone.

Overall, the creature was perhaps a meter in height,
a meter and a half long, and seventy or eighty centimeters wide.

Starkman stared at it for a long moment, taking it
in. It was not slimy, nor especially repulsive or horrible;
it was merely strange beyond anything in his experience. It looked absurdly out of place in the almost-empty
room. The peculiar computer console was clearly designed for its use; the buttons of the keyboard were
arranged ideally for its clustered tentacles to tap at,
and the holographic sphere was at the right height for
its four eyes.

It whistled and chirped at him through the orifice
on the top of its "head." He called, without turning,
"Have you got that voder set up?"

The thing would look more at home underwater, he
decided; perhaps it was from a high-pressure environment. He no longer had any doubt that it had come
from another planet.

As if answering his thought, it picked up a hose he
hadn't noticed, using one tentacle to place the nozzle
in its upper orifice, which irised down to hold it firmly.
The tentacle released the hose, then flicked at a valve.

That, Starkman realized, must be its breathing apparatus. It used no elaborate suit or helmet, so he
guessed that Earth's air was not poisonous to it, but
simply inadequate somehow.

"The voder's ready," called the pilot from the other
room.

"Good," Starkman answered. He backed out of the

room, hoping that nothing he had done had offended the alien.

He was completely convinced now that it was a genuine alien. There was nothing else it could be. That meant, he realized, that there might really be a Galactic Empire, just as he had been told, and that its motives might be just as benevolent as he had been told. They could equally well be sinister, or so inhuman as to defy comprehension.

The Imperial zombies probably *were* androids, then, he told himself.

His whole cynical theory of a human conspiracy crumbled away and vanished, leaving him at a loss in knowing what to do.

He paused in the little hallway. "Should I close the door?" he asked.

"Suit yourself, or ask him," the other man replied.

"You ask him," Starkman said.

He heard the other's voice briefly. After a pause, something flickered in the holographic sphere, at an angle and speed that made it impossible for him to see what it was. The alien tapped keys in response, its tentacles moving in a peculiar accelerating rhythm; a moment later the flat voice of a machine spoke from the living room, saying, "I am not concerned."

With a final hesitant glance, Starkman left the door open and returned to the living room.

The other man was standing by the computer holding a small metal object. "I've got it hooked up for voice input and output both, on this end, but on his end it's still keyboard and printout; he needs to breathe. I heard him talking to you, but he can't do that for more than a few seconds at a time."

"That's fine," Starkman said. "I understand." He glanced back through the open doors at the alien. It waved a tentacle, and typed something.

There was a pause. A second or two later the computer said, "I greet you politely."

"Why is it so slow?" Starkman asked.

"It takes time for the computer to translate," the other explained. "The Watcher doesn't know any English; he's reading and writing his own language, which

I can't make any sense of any more than he can speak
English. The computer has to translate both ways, a
sentence at a time, and it's not very fast. I've got my
own machine here hooked into an interface that someone in the fleet rigged up, and that's hooked into the
Watcher's machine, which does the actual translating.
With all those steps you can't expect a lot of speed,
particularly with the voder."

"Oh," Starkman said. The explanation seemed reasonable. "Hello," he called, looking at the alien and
waving.

The pilot handed him the little metal object. "Talk
into this," he said.

Starkman looked at the thing, but could not identify
it. There were several small controls on one end and a
grille in one side. "Hello," he said into the grille.

A second later something flickered in the holographic sphere.

Tentacles tapped, and Starkman waited for the reply, "I regret politely the inconvenience of this system."

"Oh, that's all right. Uh... what are you?"

"I am not assured of your want," came the reply after
a pause. "My personal name is not translatable. My
species name is not translatable. My human compatriot
calls me Watcher One. Watcher One is a faulty translation of a title I qualify to use."

Starkman found himself watching the alien, hoping
for a readable expression, but of course there was none;
the thing had nothing that he could think of as a mouth,
and its eyes were unlidded and unchanging in appearance. It did flex its various appendages, but Starkman
could make nothing of that. Three eyes were usually
directed at the computer read-out in the sphere, while
the fourth watched him through the hallway.

Realizing that watching it only served to confuse
him, Starkman stepped back out of sight, and after a
moment's hesitation settled onto a couch. "Excuse me,"
he muttered.

"You require no pardon," the alien replied. "Politeness must be transformed between species not previously in contact."

There was a brief pause, but before Starkman could say anything, the computer's voice spoke again.

"You are concerned with many questions. You wish to comprehend events. I will attempt to explain events. When you have questions you will ask them without regret. Will you communicate through this system?"

"I think so. I mean, that sounds all right."

He glanced at the pilot while waiting for the response; the man was leaning casually against a wall, ignoring the conversation.

"The machine that translates is not perfect. I ask politely that you attempt to communicate clearly."

Stamping down the urge to reply with "Sure" or "Got it," Starkman said, "I will."

There was another pause. Finally, the alien said, "I am called Watcher One because I am the most highly responsible member on your planet of an organization called 'the Watch.' A moderate number of members of the Watch are among the members of the expeditionary force. You have the potential to fear that I am a member of the government. The Watch opposes purposes that the government pursues. Do you understand?"

"I think so. The Watch is a sort of opposition party in the fleet."

"That is not clearly correct. Loyal members of the government are not cognizant of the existence of the Watch. The Watch takes actions that serve to delay government actions. Members of the Watch appear to be loyal members of the government while they take actions that interfere with purposes the government pursues. Do you understand more clearly?"

"Do you mean you're an underground? You're subversives?"

"I believe that you comprehend correctly."

Starkman glanced at the pilot; he nodded.

"You're the head of the underground on Earth?"

"Yes."

"Are you allied with the other Underground, the group that I was with earlier?"

"No."

"Why not?"

"The members of the Underground act violently. Members of the Watch do not act violently."

"I'm not sure I understand."

"The members of the Underground pursue their purpose inefficiently and with no concern for harm to sentient beings that may result from their actions. Members of the Watch do not perform actions that may result in harm to sentient beings. Harm to sentient beings is forbidden."

"That sort of limits your effectiveness, doesn't it?"

"We are more concerned with acting correctly than with acting efficiently."

"That's very noble, I suppose."

There was a longer pause than usual before the alien replied, "Ask questions."

Starkman glanced at the pilot, who did nothing, then looked quickly around the room. He saw nothing that gave him any great ideas.

"Are you aliens planning to colonize Earth?" he asked at last.

"No. The Watch does not approve of colonization. The government on your planet has other purposes and has no interest in colonization."

"All right, you Watchers are too moral to do us any harm. What about the government? Are they planning to use people for slaves, or cattle?"

"No. Machines are more efficient servitors than humans. Intelligent life is not used for food or other direct physical utilization."

"Then what *do* they want here?"

"A complex explanation is required. I will attempt to accomplish the required explanation. A single purpose has caused all actions here. The purpose has caused the alteration of your planet's climate, and the usurpation of your planet's government, and the pursuit of yourself. Will you listen patiently?"

"Go ahead."

"The representatives of the Governor told you that they are representatives of the government of the Galactic Empire. They are not. They are citizens of the Galactic Empire. That is true. The Galactic Empire exists. The name is not correct. The Empire controls a

small portion of the galaxy in the region approximately
in correspondence with the constellation of Sagittarius,
and areas farther in toward the galactic core that are
not visible from your planet. The government of the
Empire is an oligarchy. I am not assured the translation
is clearly correct. A small group of individual beings
controls the government. The beings are not selected
by any rational process. The expeditionary force on your
planet is controlled by a member of a group of beings
who wish to supplant the beings who control the government. The beings of the group have no official standing in the government. They have attempted to obtain
control of the Empire by force. The beings in control of
the Empire have opposed them by force. The result is
a continuous conflict." There was a brief pause. "Do you
understand?"

"I think so," Starkman said. "You mean that Earth
has been taken over by a rebel faction of the Empire,
rebels who are waging a civil war against the established government."

"Your statement is not clearly translatable in complete form. I believe you understand correctly. I will
proceed."

Starkman nodded and said, "Go ahead," before he
realized that the alien had not stopped speaking.

"Do you understand the nature of interstellar travel?"
it asked.

"What do you mean? I don't know how a star drive
works, if that's what you mean."

"I ask the question in another method. Do you understand the distances between stars?"

"I'm still not sure what you mean; I know that the
stars are light-years apart and that it would take years
and years to travel from one to the next without some
kind of faster-than-light drive."

"You understand. Velocity is constrained."

"Velocity is constrained? I *don't* understand."

"It is not possible to travel at velocity greater than
the velocity of electromagnetic radiation."

"I know it isn't ordinarily possible, but you must
have traveled faster than light to get to Earth." A sud-

den thought burst into his head, and even before he heard the answer he knew what it would be.

"No. It is not possible to travel at velocity greater than the velocity of electromagnetic radiation."

That was a depressing piece of news; Starkman had assumed that the aliens possessed some way of circumventing the laws of Einsteinian physics, and that mankind might reach the stars with it. It appeared that he was wrong; no species as short-lived and fragile as humanity would do much star-hopping at sublight speeds. He said nothing, and after a few seconds the alien continued.

"Beings who do not naturally die control the Empire. The extent of the Empire makes necessary government composed of undying beings. Do you understand?"

That made sense. There was no reason to assume that in all the galaxy every intelligent species faced death from old age. Starkman answered, "Yes."

The concept was food for thought. It was indeed obvious that a galactic empire could not be run by people who needed half a lifetime to get anywhere. He asked, before the alien's next sentence came through the translator, "Are there many such species? Are you one?"

"The majority of intelligent species have limited lifespans. The Empire is controlled by the very small minority. I am undying."

"Oh. How old are you?"

"The question is trivial. The question is irrelevant. I cannot answer. I have not maintained a record. The measurement of time is not universally consistent."

"All right, I'm sorry I asked; go on with what you were saying."

"You will understand that undying beings are not concerned with efficiency. Undying beings are not limited in time. The beings in control of the Empire and the beings who oppose them and wish to obtain control are all undying beings. Therefore they have prolonged the conflict without limit. The conflict has continued for a very large amount of time."

"How big a war is this? I mean, if these guys have been shooting at each other for a few hundred years, they must have done a lot of damage."

"They have done a large amount of damage. You have spoken truly. The conflict is not constant. Times occur when no conflict takes place."

"It's sort of off-and-on?"

"Yes. You understand. Communicating through a translating device is cumbersome, is it not?"

"Yes, it is. Go on; what has this galactic civil war got to do with me?"

"A long time in the past, an event occurred. Researchers who were controlled by the beings who oppose the beings who currently—"

Starkman cut the alien off. "Call the two groups the Empire and the Rebels."

The pilot smiled as he leaned against the wall; he recognized the cinematic reference.

"Yes. The researchers were controlled by the Rebels. I politely acknowledge the alternative terms to be more convenient. The researchers were in a ship. Ships are more difficult to locate and destroy than structures on planetary surfaces."

"I can see that."

"A researcher made a new thing. The thing was an unlikely happenstance. It was not planned. Beings have attempted to repeat the event for a very long time. They have not repeated the event."

"Someone hit on a fluke, a chance discovery; right. So what?"

"The researcher had made a thing that explodes stars. It creates supernovae. Do you understand?"

"I think so; this guy found a way to make stars blow themselves up."

"You understand. No being had done this thing before. No being has done this thing in the long time since. Supernovae can be a very great weapon. The Rebels can destroy the government and control the Empire with supernovae."

"I understand that, but if this happened a long time ago, why haven't the Rebels won their war?"

"The information is not in the possession of the Rebels. An Imperial warship located the researchers' ship and pursued it and destroyed it."

"Well, if it was destroyed, then what's the point of

this whole story?" Starkman was becoming exasperated with the awkwardness of the alien's speech.

"D'you want me to take over?" the pilot asked. "I know most of it, and anything that's not clear you can ask the Watcher."

"Well..." Starkman considered. It was fascinating, in a way, to listen to the alien, but it was also slow and clumsy using the translator, and the result was not always very clear. "Yes, I think that'd probably be for the best." He passed the microphone pack to the other man's outstretched hand.

"Hello, Watcher? This is your Watchman. I will tell him more. The translating machine is not convenient. Is it acceptable?"

"Yes."

"Good." He put the device down on the computer cabinet, leaned back against the wall again, and began, "The reason that the story about the research ship is important is because there were survivors. Most were lost for good, and presumably died long ago; nobody seems to know how many actually got away. Maybe there were only a few to begin with. At any rate, only one is of any importance, so far as we know. A genetic engineer, who had been on the research ship primarily to produce test subjects for the other scientists, got away in a small ship. He wasn't much of a pilot, and the ship itself wasn't very intelligent, being just a lifeboat, so he wasn't able to do much real navigation. He certainly had no chance of getting back to anywhere civilized, by his standards; the ship wasn't self-supporting, and his supplies were very limited. He did find a planet that was habitable for his species in a neighboring system, and he put down on it—except that, as I said, he wasn't much of a pilot, and he crashed. To be specific, he crashed on Earth, probably in what's now the Sahara Desert— or rather, what *was* the Sahara Desert, and is now the Sahara Development Area."

"He did? When? Even in the Sahara, I would think an alien spaceship would be found eventually."

The other signed. "You didn't understand what the Watcher meant when he said that this happened a very long time ago. I'm not sure of the exact time myself, but it was at least forty thousand years ago."

"Oh." Starkman felt appropriately squelched; he had been thinking in terms of four or five centuries.

"After the crash," the Watchman went on, "the alien's ship was hopeless; his computer was dead, for all intents and purposes, and the ship was no longer airtight, so there was no way he would ever get it into space again. His transmitter had survived, however. I'm not sure just what kind of transmitter it *was;* I'm afraid the Watcher isn't a scientist or an engineer, and we have our own communication problems, so if he knows, he hasn't been able to explain it to me. One thing I do know is that it wasn't using tachyons or anything else in that category; the message went out at the speed of light, no faster, which meant that it was strictly a monologue, not a conversation. Those transmissions were made in Rebel code—I'd been calling them the Opposition, but 'Rebel' is a better name. They were eventually picked up and recorded, several thousand years later; we're a *long* way from the Galactic Empire out here, and at that time the Rebels were mostly working their schemes out somewhere on the far side— or at any rate not the nearest side. They don't care much about time, as the Watcher said. Neither did the castaway, really; he was pretty much immortal himself. He knew that it would be at least centuries and probably more before anyone heard him, but he still had hopes of being rescued by whoever heard him. Because of that, he said that although he had with him all the data on the method of triggering supernovas, he wasn't going to transmit it; he wanted his comrades to come and get him off our little mudball, and he didn't think they would if he didn't give them a damn good reason. The Watcher tells me he was probably right; the Rebels aren't too big on such things as loyalty and compassion."

"Is the Empire any better?"

"Probably not."

"I think I see where you're leading; these aliens who have taken over Earth came here looking for their lost comrade and the formula for making stars go blooey.

What I don't understand is what that has to do with me."

"I'm getting to that. If you're interested, I could just play some of the taped messages for you—in translation, of course. I've got them all here in the computer, I think."

"You do?"

"Yeah, sure; let me show you." He reached over and punched buttons; the lights dimmed and died, and the video screen glowed softly blue.

The blue faded into black, and for a moment the only light in the room came from the glow of the city outside the curtained windows. Then the screen lit, showing a blurry expanse of desolate grassland.

"It was a real bitch getting an image," the Watchman said in a confidential tone. "They use a spiral scan, instead of our parallel lines."

There was a crackle, and a voice spoke haltingly from the computer. "I am here," it said. "I show you the star." The view moved upward to the sun, and the screen grew uncomfortably bright. "I will show the other stars later."

"He started every message like that; he usually forgot to include the stars, or maybe he just couldn't get a decent picture," the pilot said.

"I have made food by genetically altering the indigenous life. I can live indefinitely. I will wait for rescue. I have knowledge of the work of…" There was a moment of silence.

"The researcher's name doesn't translate," the Watchman explained.

The voice began to recite an explanation of how the ship had crashed; the pilot punched a button. "You're not interested in that."

The image faded out, then returned, still blurry. Now it showed a tree and banks of what might have been a stream.

"I am here," the voice said. "I have located sentient indigenous life. The natives are very primitive. They are sentient. They are not civilized." A figure walked into the visible area; it was human, but seemed oddly wrong to Starkman. It was a short, stocky man with a low, wide forehead. He had no beard, but seemed very hairy otherwise. He was naked. Starkman was unsure whether he was a heavily tanned white or one of the lighter-skinned varieties of blacks; he did not look Oriental, though it was hard to be sure of much of anything. The picture quality was poor and gradually deteriorating as he watched; details were lost in the fuzz and flicker.

Then the picture blurred away completely. "They lost the transmission on that one," the pilot remarked. He pushed a few buttons. "The next few are pretty boring; he goes on to tell about a few of his discoveries, such as that the local chemistry is DNA-based and that the natives use tools. Let me skip ahead."

The screen went white, then blue, and then a new image appeared. This was an interior scene, of a room cluttered with mysterious apparatus, presumably aboard the wrecked ship. The image was dim and splotchy, with a wavering gray stripe running diagonally across it.

"I have experimented. I have analyzed the DNA of the native sentients. I have altered it. I have made servants. They will help me to survive. They are not stronger than the true natives. The diet would not permit. They are more vicious. They will help me to survive. You must come. I wait." Something moved in the image, and a boy stepped into view. He was perhaps six or seven years old, brown-skinned, black-haired, but without the wrongness in his features that Starkman had noticed in the earlier image. A girl came into view beside him. "They are my servants," the voice of the stranded alien said through the translator. "I am here. You must come. I wait." The picture vanished abruptly.

"You see what he did?" the Watchman asked.

"I think so," Starkman answered. "Those servants of his—they must have survived and outlasted the ordinary people. They're like us." The thought that he had just seen his distant ancestors was slightly unsettling.

"Cro-Magnon, that's what they are. I think the original natives were Neanderthal. And we all know that Cro-Magnon man survived and Neanderthal didn't." He pushed buttons again. "That's interesting enough, but this next message should really grab you."

The image of the cluttered chamber returned, but more dimly, and with the diagonal stripe gradually shifting angle toward the vertical.

"I am here. I am not healthy. I am loyal. I want to preserve the knowledge. You will have the knowledge. I may die. My death will not destroy the knowledge. I have put it in binary code. I have put the code in DNA. I have put the DNA in an intron. I have linked the intron to a distinctive trait. I have put the intron and the distinctive trait in one of my servants." A telescoping limb with horny protrusions here and there appeared at the edge of the image, holding a newborn human infant. "This is the servant. This being has the knowledge in its genetic material. You will see its eyes. The color of the eyes is the distinctive trait. It is a simple dominant. I wait. You will come. I am not healthy. This being is very healthy. It is strong and intelligent. It will be fertile. I have made it superior. It will survive. The knowledge is in its genetic material. It will be preserved. Its offspring will have the knowledge in their genetic material. It will be preserved."

The baby, cradled in the curve of the alien appendage, woke; his eyes opened, and he looked at the camera. The image flickered and vanished, but not before Starkman glimpsed, through the blurring and snow, the infant's yellow eyes.

Chapter Nine

"That was the last message," the Watchman said. "Apparently the transmitter failed for good right about then, and the alien probably died not very much later. He certainly didn't live long enough to turn up in recorded history."

"What about the baby?"

"We don't know. He was apparently pretty sure it would survive, and his other servants obviously did, or we'd all have brow-ridges today. He figured that putting the data in the kid's genes was the best way to preserve it; he must have known that his ship wouldn't last forever, and that any record he might make would wear away eventually."

"That word he used—what's an intron?"

"It's a blank space on a DNA molecule, sort of. DNA is made up of long strings of four basic compounds— adenine, thymine, guanine, and cytosine—arranged in pairs, adenine with thymine and guanine with cytosine. They can be either way around, so it's a quaternary system, base-four, not really a binary system, but it's perfect for encoding data—which is, after all, just what DNA does. Except that our DNA evolved naturally, or at least most of it, except for whatever that alien changed; it wasn't designed by a computer expert, so it's not very efficient. There are long sections that don't do anything, that are apparently just random groupings. The sections that do carry information have trigger groupings at each end that tell the other compounds to interact with them, and then the links in the active part are arranged in groups of three that synthesize all

THE CHROMOSOMAL CODE 137

the different proteins we need. The stretches on the other side of these groups, the pieces that carry no information, are called introns. If you were able to manipulate DNA the way our castaway could, it'd be simple to encode anything you want in a long binary pattern— say, guanine-cytosine and cytosine-guanine, because that bond's stronger than the adenine-thymine pairing—and stick it into an intron, and locate it right next to the trigger group for the gene for yellow eyes. The connection might get broken eventually, or the coding broken up, but it'd take several generations. You understand?"

"I think so. This creature figured that when his friends came to rescue him, even if he was dead they'd find people with yellow eyes and be able to analyze their DNA for the information."

"Exactly."

"And they came and they didn't find anyone with yellow eyes."

"Until you came along."

"Until I came along. If they knew they were looking for yellow eyes, why did they get all excited about an albino? The doctor from the Welcome Center mentioned that. And why do they take blood samples from everyone?"

"You saw how poor the image was, didn't you?"

"Yeah, so what?"

"That image is a computer-enhanced idealization of a message transmitted over a couple of thousand light-years. It was recorded and analyzed by aliens who had never seen Earth. It's not color-corrected; the aliens had no way of knowing how Earth was supposed to look. That kid's eyes could have been anything from infrared to pale green—yellow is just the most likely. When they first got here, they thought the trait they wanted was *green* eyes; they didn't see any yellow ones, and they figured the color was off. Except that apparently people with green eyes haven't got anything coded in their DNA, quite aside from having white eyeballs. When they realized that wasn't it, they started looking for red and yellow and orange. Meanwhile, they took blood samples from everybody so that no one would know they were sorting by eye color. If they get desperate enough, they may start analyzing every single sample, hoping they find the data even when the trait isn't visible."

"Okay, so the Rebels got these messages and sent out a fleet and it arrived here in 1999. Right?"

"No."

"No?"

"No. They arrived in 1946. It took them that long because the war was going badly when the message finally reached them, and they had a lot of trouble figuring out exactly what star they wanted—as I said, the messages weren't color-corrected, so that they couldn't get very much use out of analyzing the sun's spectrum. And then it took a while for the Governor to put together a fleet and slip it out of the Empire without being noticed by the Imperial government. Still, they got here in 1946, as nearly as I can figure. They did a search, and found no trace of their castaway; he was obviously a long time dead. They had no trouble finding people, though, so they kidnapped a few at random. As I said, when they didn't find yellow eyes they tried green."

"They kidnapped people? In 1946? Why didn't anyone notice? Why didn't anyone see their ships?"

"What makes you think no one saw them? They were careful, but they were spotted repeatedly between 1946 and now. They even shot down the occasional plane that got too close."

"They did?"

"Their scout ships are saucer-shaped, some of them; does that ring any bells?"

"Oh."

"Of course, no one believed the reports—or at least no one important. And no one spotted the main fleet because it stayed out of sight; they've got antidetection devices so that anything we might do would be like Nazi Germany trying to spot Stealth bombers."

"Hey, if humans were invented here on Earth by a genetic engineer, where did the zombies come from? I mean the people from the fleet; are they androids?"

"I'm getting to that. You see, when they couldn't find any trace of anyone with the information they wanted, they decided that they'd just have to take over openly and sort through the entire population one by one, analyzing blood samples and looking for yellow eyes. They didn't want to have to shoot their way in, though; fusion bombs are more than they came prepared to handle. So they took some debris from somewhere—asteroids, or a comet, or somewhere, I'm not sure—and made a dust cloud out of it, between Earth and the sun. I don't know the technique exactly—I'm not an astrophysicist—but somewhere they laid down this dust screen that cut off a lot of the sun's energy before it could reach us. That caused the ice age, of course, and drove just about the entire population into the equatorial regions, where there was a lot less territory to cover. And of course it disrupted all the governments very nicely; when they arrived nobody was in a position to put up much resistance, and things were in so bad a mess that help from anywhere was welcome, even invaders from outer space as long as they could feed people. It was a pretty slick maneuver, actually—the Watcher tells me it's the standard method for dealing with civilizations at a certain level, invented by some military genius a few millennia back."

"You still haven't explained their pseudohumans."

"Oh, yes. Those are modified clones. Remember, they'd been kidnapping and studying people for about half a century, and they knew that we're pretty xenophobic. They didn't have any humanoids along—apparently there aren't any that live long enough to do much interstellar traveling—but they needed humanoid ambassadors and so forth. They had long ago developed their translating computers for every major terrestrial language, while they were hiding nearby and kidnapping people, but they needed representatives who could go out and be seen in public without causing riots. They didn't have any, so they made some. It wasn't that hard; they took a few sample cells from some of the people they kidnapped and grew clones. They were in a hurry, so they accelerated their growth as much as they could without producing complete morons; they got what you call zombies. It takes them about eight years to grow to adult size—but in some

regards they don't develop any faster than anyone else.
The oldest ones, their original experimental ambassadors, are about twenty now, but most of them are much
younger; the ones they use as cannon fodder, working
as guards or search parties, are generally no older than
nine or ten. They'll probably be dead at thirty, by the
way, so don't be too hard on them. Anyway, with the
accelerated growth, they tend to be erratic and emotionally immature—though not much worse than most
eight-year-olds. They're all male because they didn't
want any emotional complications resulting from sex-
ual entanglements between clones—and in fact they've
got pheromones designed to make them slightly repulsive, as if their behavior weren't enough, to keep them
from getting entangled with human women. They're all
sterile, so you don't have to worry about half-breeds.
They picked males for their one sex because they knew
most human societies are male-dominated, and they're
all white because Earth seemed to be white-dominated.
Besides, for some reason most of their kidnap victims
were white; I don't know why. Maybe it's because whites
have more varied eye colors. Once they had their clones,
they used those as their representatives, and made it
appear they were the aliens. Actually, no clone has ever
been any farther from Earth than the moon. They admitted they had intelligent machines, since we had our
own moronic ones, and when an alien has to work with
humans he usually does so disguised as a machine."

"What about the Governor?"

"He's a real alien, the commander of the whole expeditionary force. In fact, he's a high official in the
Rebel command structure, and the whole thing was his
idea. They decided using a figurehead would be too
much trouble, and that an aura of mystery would help
keep people in line."

"I see."

"So they took over, and set up their Welcome Centers
and health checks, and began doing genetic analysis of
the white corpuscles in blood samples whenever they
hit anyone with some hereditary peculiarity, particularly unusual eye color. They went through the entire
population of the warm countries without finding any-

one who carried the formula for supernovas. Finally they got desperate and began searching the cold countries as well, bringing in stragglers."

"And they found me."

"Exactly. You, Mr. Starkman, are the only person who might be what they're looking for—and, by a great stroke of luck, they haven't got you."

Starkman leaned back and stared at the ceiling. "Where do you and the Watcher fit in all this?"

"That's simple enough; we're members of the Watch, and the Watch is dedicated to peace. It's an organization of sentient beings of hundreds of different species that has infiltrated every level of both the Rebel organization and the current government of the Galactic Empire. We're all working, by peaceful, nonviolent means, to end the ongoing civil war and to reduce conflict in general. We don't want *anybody,* Rebels or Empire or even humans, to know how to trigger supernovas. It's just too damn dangerous a piece of knowledge to have around; sooner or later someone would use it—probably the Rebels would use it as soon as they got it. That's certainly what they plan; they think it would swing the balance of power their way, and they're ruthless enough to use anything they can. It wouldn't bother their high command at all to wipe out a few inhabited systems, the callous bastards. Look how casually they ruined Earth!"

"So what are you doing about it?"

"Whatever we can. The Watch has suppressed research, destroyed data, delayed the search for Earth, prevented battles, obstructed conquests, and done everything they could to wind down the war and prevent either side from developing any new superweapon, including the supernova bomb. They've reduced the destruction immeasurably already; they might end the war in another few centuries, because the Rebels are already declining significantly in power and prestige. That's one reason that the Governor and his people are looking so desperately for the supernova bomb; it may be the only thing left that can give them a good chance of winning."

"I don't see that they've done much here on Earth;

it looks to me as if the Rebels have been doing just about anything they want."

"We're trying, dammit! It's not easy, you know, when you're sworn not to use violence. A few quick assassinations, or even just one, might change everything, but the Watchers are all sworn, when they join, never to knowingly harm any sentient intelligence. And security was very tight on this mission, the Watcher tells me; the Rebels weren't worried about the Watch—I don't think they even know it exists—but they were scared of Imperial spies. There are only a very few Watchers in the expeditionary force, and most of those are in positions of little power. Watcher One is first in command in the Watch here on Earth, but in the expeditionary force as a whole he's just an assistant quartermaster, as near as I can interpret his duties."

"All right, so there aren't many of them, but there are obviously a few, and they can recruit Earthmen, the way they must have recruited you. Maybe they can't do much, but they must be doing *something!*"

"Of course they are! *We* are! You know that the Rebels have been handing out their technology like candy, to keep the natives quiet and happy; well, the Watch has been making sure that they give out more than they intend to, and that people are finding out just how everything they receive works. Not just techniques for growing better grain, but the whole theory of antigravity, for example. The Rebels gave out the aircars, but they never thought mere humans could understand how they worked. With a bit of coaching from the Watchmen like myself, human scientists *did* figure it out. Terrans were already beginning to develop decent computers, and the Watch has been training human computer experts in the design of the best the Empire's got; the Rebels' information network was put in command of all the computer and communication networks here on Earth, what people call 'Max,' but we've managed to reverse that and tapped into command of their network without their knowledge. At least, we think we have; sometimes it's hard to say just who or what controls any given computer. We've even been disseminating the plans for their weapons—we know that

Earth people will build them, which is bad, but it's the only way that they can learn to defend against them, or to disable them when the time comes to take back the planet."

"When will that be?"

"I don't know. Not yet; not while the Governor is still so firmly entrenched."

"All right, so you're doing your best. And you've brought me here to keep me out of the government's hands, because the government of Earth is run by the Rebels and I've got the information they need to blow up stars programmed into my genes."

"That's right."

"You will understand, I hope, that this is all a bit much for me to take in all at once. A week ago I was living alone in the snow, thinking my only problem was finding food and dealing with Mother Nature. I didn't know that there were aliens anywhere in the universe, let alone three different factions of them here on Earth."

"Only two, we hope. We don't think that any Imperial spies got in."

"Whatever." Starkman waved the correction away. "And we're all descended from a bunch of genetically engineered servants some shipwrecked alien whipped up to make his life a bit easier."

"Right."

"And I'm apparently the only known descendant of the one he used as a data bank."

"So far as we know, yes."

"So you've gotten me away from the government; wonderful. Now what happens? Do you hide me here in your apartment until I die of old age?"

"No. At least, I don't think so."

"What do you think, then?"

"I think I'll let the Watcher explain what we had in mind." He picked up the microphone device and said, "Watcher, have you heard what has been said?"

"I have heard."

"Mr. Starkman wants to know what we want him to do."

"I have heard. The Watch may not kill. Do you understand?"

"Yes, of course," Starkman answered. "Your Watchman explained that."

"We prefer that the information in your genetic material be destroyed. Do you understand that?"

"I understand that you want that, but you can't do it without killing me, can you? You'd have to take every single cell and cut out part of a chromosome. You can't do that."

"We cannot. You are correct. We prefer that the information be destroyed. We prefer that you be destroyed. We may not kill you. You may kill yourself. We ask you to kill yourself. Your death may prevent a very large number of deaths."

"What?" Starkman was shocked at the suggestion.

"If you kill yourself we will destroy your body. The information will be destroyed."

"The hell it will! Ten minutes after I'm dead some guy with yellow eyes may turn up in Outer Mongolia somewhere!"

"We would ask him to kill himself."

"Sure, if you got him away from the government in time! I'm not about to kill myself for you, whoever you are. I can't be sure that any of your story is true, after all. Oh, you tell it pretty convincingly, but I don't *know* that it's true! And I'm not convinced that I'm really the long-lost message carrier; if the original guy was so wonderful, such a great survivor and breeder, why aren't there millions of people with yellow eyes? Why just me, out of five billion people? It doesn't make sense!"

"It is true."

"Well, then maybe the gene has mutated! Maybe the information's no good!"

"You may be correct. We do not think that the possibility is likely."

"After forty thousand years?"

"The genetic engineer was very competent. He would make his carrier very stable."

"Maybe so, but look here, Watcher, I'm not going to kill myself for you or for anyone. So now what?"

"We considered the possibilities. We will hide you."

The human Watchman said, before Starkman could protest, "Not here, don't worry. We'll find a place for you somewhere where you can run your own life, if that's what you want."

"Why don't you just take me back to Pennsylvania?"

"Because the Rebels would find you again if we did. No, we've got to put you someplace where they wouldn't think of looking for you, someplace they think everyone's accounted for."

"Where?"

"Well, as far from here as we can manage. Have you got any friends or relatives who might shelter you?"

"I had a sister once; I don't know where she is."

"No, I don't think that would do; they'd check with her if she's your only relative. What about friends?"

"It's hard for a freak to make friends," Starkman answered bitterly. "Any I might have had once are probably dead, and if they aren't they've probably forgotten I ever existed."

"We'll just have to do what we can, then. We can drop you off in New Denver, if you like; that's an American colony in the Andes that's accepting new arrivals."

"Wonderful. The first person who sees me will call the cops."

"No, he won't. We'll help you blend in—get rid of that coat, clean you up, cut your hair, and give you some decent modern clothes, give you a few lessons in how things work, loan you a few units—that's the money now, units."

"I appreciate that, but what are you going to do about my eyes? Anyone who sees my face will turn me in."

"I know. There's not much we can do about that; we don't have any contact lenses on hand that can do the job, I'm afraid. I'll see if we can't locate some, or have some made, but it'll be tricky, having to color both the white and the iris. I don't think they'll be looking for you in New Denver; they'll assume you're still here in Capital. For now, just keep your sunglasses on."

Starkman shrugged. "I always have," he answered.

Several hours later and several hundred kilometers away, the aircar slowed to a stop, hovering a few centimeters off the ground. Starkman gathered together his gear and clambered out onto the empty plaza.

The eastern sky was red with the coming dawn, and high overhead the top of the city dome sparkled in the first rays of the sun. The streets and buildings that stretched away to the west looked slightly unreal in the dim light and their clean bright newness.

"Welcome to New Denver," the Watchman said. "You shouldn't have much trouble in finding a place; I hear that the city's still about a quarter empty. It's one of the government's least successful projects."

"Right. You told me that already."

"Just reminding you."

"Thanks."

"I don't think they'll think to look for you here."

"I hope not."

"Got any final questions?"

"No. I'll manage."

"I'm not so sure about that. Look, I've got something I want to give you, in case you change your mind."

"What's that?" Starkman paused, clutching the shoulder bag the Watchman had provided him with, and looked at the two little objects that gleamed on the man's palm.

"The bullet-shaped one is a beeper, a locating device; if you ever decide you need to get hold of the Watch, and you can't use the comnet code I gave you, push down on the top and turn it ninety degrees. Someone will hear it and come get you."

Starkman looked at it, and shook his head. "No. Keep it. If I get in trouble I'll get myself out or not without dragging you people in. You probably couldn't get there in time to do any good; it'd just compromise your organization. And if I'm not in trouble, I'll just phone— use the comnet, whatever you want to call it."

"All right, I won't insist. You may be right."

"What's the other one?"

"I was about to tell you. If you change your mind and decide that the Watcher was right, or if you're ever in a spot where you can't possibly get away, you may want it. It's a flashbomb."

"A what?"

"A flashbomb. It will vaporize anything within a radius of about a meter and a half—silently, too, though there is a bright flash. You can use it to suicide, and be sure that there won't be enough left for genetic analysis."

Starkman stared at the little thing, shaped like a disposable cigarette lighter. The form struck him as slightly odd, since he hadn't seen any sign of tobacco in use in the past decade; he suspected the weed was extinct. "How does it work?" he asked.

"There are two ways you can set it off; press this down," he said, indicating a tiny metal tab just below the nozzle, "And then flick the wheel, and it'll go off immediately. Or turn this," indicating the control that would ordinarily have adjusted the size of the flame, "and then flick it, and you've got a timed explosion. It's not calibrated very accurately, but if you turn the dial as far as it will go, you've got about thirty seconds after you flick the wheel."

Starkman accepted the little gadget and turned it over in his hands. "Where'd you get it?" he asked.

"The Watch gives them to all their Terran recruits, just in case, so they can destroy evidence or suicide if necessary."

"I thought they didn't believe in violence."

"We don't."

"Suicide by bomb-burst isn't violent?"

"Maybe the method seems violent to you, but suicide is the right of any sentient being, and something like this leaves no traces that might give away secrets. It's a small confined explosion, and the timer is designed to let you get away from anyone else—or to get clear if you're destroying something, like computer memory or papers."

Starkman stuck the flashbomb in his pocket. "Thanks," he said. "I don't plan on suicide, but I might find a use for this."

The Watchman looked worried. "It's not a weapon," he said. "We don't want to promote violence."

"Don't worry about it," Starkman said. "I know you're all sworn not to harm anyone." He stepped back, waved, then turned and began walking down the nearest street.

Behind him the Watchman watched him go, then

turned the aircar around and headed it back out through the city airlock.

Starkman paused as he heard the automatic lock-door cycle, and turned back to be sure that the Watchman was gone. He pulled the flashbomb out of his pocket, looked it over carefully, then returned it. "Maybe they're sworn not to kill," he muttered to himself, "but I'm not."

Chapter Ten

Starkman had no very clear idea of where New Denver was, save that it was atop a peak in the Andes. The mountaintop had been sliced off flat and the city built in its place, with the dome to keep the air at a comfortable pressure and temperature. The buildings were clean and new and very boring, row after row of neat concrete-and-glass boxes on straight numbered streets and avenues, like the ideal city of fifty years earlier—an ideal that had proven unlivable. The Watchman had told him that it was supported by hydroponic farms that took advantage of the increased exposure to ultraviolet at high altitudes, but had no real industry to speak of; transportation of raw materials would require antigravity, and that wasn't yet cheap and widespread enough.

He himself had reached it by a long and uncomfortable ride in the aircar, a ride that wound up and down and back and forth, looping recklessly through the dying jungles dodging trees, soaring over mountains, following rivers, doubling back on itself, and in general designed to be untraceable even with the best equipment the Governor could produce.

Toward the end of the ride he had very much regretted giving the Watchman his old coat and flannel shirt; the open car had been cold and windy, and the low-necked, short-sleeved shirt he now wore—which was made of something akin to very thin lavender velvet—was not particularly warm. He knew that he was not actually any colder than he had been many times before, while living in the wastelands, but he was no longer accustomed to it, having been spoiled by his brief stay in warmer climates.

Now that he was inside the dome he was warming rapidly, aided by the brisk walk he was taking down the street, looking for a place to stay.

Gold lettering in a window caught his eye, and he read: RENTAL AGENCY. That was exactly what he was looking for.

The office was decorated in red and orange and reminded him of an old-fashioned fast-food joint, and the young man at the red-and-orange desk was delighted to help him. The agency, he was told, could provide just what he was looking for—whatever it was.

"An efficiency apartment, six-month lease," Starkman said.

"Ah, another new arrival!"

"What?"

"You're newly arrived from up north, aren't you? I can always tell. Where did you straggle?"

"What makes you think I'm new?"

"Nobody calls them efficiencies any more, sir; they're just called 'standards.'"

"I'm just old-fashioned, all right? I've been staying with friends in Capital, not talking to rental agents."

"Whatever you say. We have plenty of standard units, sir; what would you like?"

"What are the choices?"

"Oh, what floor you'd prefer, and color scheme, and service. Oh, did you want furnished or unfurnished?"

"Furnished, near the ground, and I'm not picky about color, but nothing too bright."

The clerk punched buttons on a keyboard set in his desk, then asked, "Would brown and tan suit you? And what about service?"

"Brown's fine. What kind of service? I don't need a maid, if that's what you mean."

"No, of course not; I mean the comnet. What sort of Max do you want?"

"I don't want maximum anything."

"No, sir, not maximum, Max. That's the name for the central computer."

"Oh." Starkman suddenly remembered events in Dr.

Curtis' examining room, and other mentions of the computer network. He mentally chastised himself for not realizing what the clerk was talking about, but reminded himself that he had been a long time without sleep. "I don't need any, thanks."

"No phone?" The clerk was audibly surprised.

"I'll want a phone, but no computer."

"Sir, the phone system is *part* of the computer; you can't have one without the other."

"Oh."

"I think you'd better take it step by step, if you don't mind."

"Maybe I should," Starkman admitted.

Five minutes later he had signed himself up for phone service, audio only, and small-screen video with news and bulletin service. He had turned down a bewildering array of other options, including hardware from keyboards to voders and holographic video walls, and software from simple four-function calculation to interactive theater.

He wanted to keep his contact with Max to an absolute minimum. He had no clear idea of who controlled what aspect of the comnet, and he wanted as little data about himself as possible to be accessible. He was wary of anything that allowed for input from his end; once he realized that phone service was inextricably linked to the network as a whole, he even doubted the wisdom of having that. The lines were, he was sure, something that the government would monitor carefully. He intended to use the phone as little as possible.

When he had that settled, the clerk smiled, turned the form around, and held out a pen. "Sign here," he said.

That caused an instant's hesitation; Starkman had settled on a cover identity, but he was not fully convinced he could get away with it. The society he was attempting to enter seemed to be far more complex and interrelated than he liked, thanks to Max; he was unsure just what the authorities were capable of, and whether they might be able to detect his little fraud. Would the use of a fictitious name be checked somewhere against some master memory bank? Would the government check it out when it was spotted?

That, he decided, was surely paranoid. The government had a world to run; it couldn't take the trouble to check out every single rental record, or to investigate every adulterous spouse or incognito celebrity who registered somewhere under a false name. He took the pen and signed himself "John Stanford."

The clerk smiled again, accepted the paper, and stuck it into a slot in the desk. The faint chittering of machinery sounded, and the form slid down out of sight.

When it had completely vanished a screen in the wall beside the desk lit; Starkman had not noticed it previously, as it had been red when unlit, and had blended in with the decor. It stayed red, but now it glowed, and yellow letters spelled out several addresses, followed by code symbols.

"There's only one exactly as you wanted, Mr. Stanford," the clerk said. "Those others all have more service. Of course, we could take it out by tomorrow, but I had the impression you wanted immediate occupancy, and I'm sure you'd rather not have our service people intruding on your privacy."

"That's fine. Where is it?"

"Just up the street. It's on the third floor, facing a courtyard—"

"I'll take it. Give me the key."

"You don't want me to show it to you?"

"I can find it myself, thanks."

"Ah ... I'll need a deposit."

"How much?" Starkman reached for the wad of bills in his shoulder pack.

"Twenty-five hundred units."

The amount dismayed him, but he counted it out. It took a large chunk out of his borrowed funds. The clerk accepted it, punched buttons, and fed the money into a slot—not the one in the desk top that had taken the form, but one out of sight behind the desk.

There was a click and a whir, and a door in the wall slid open, revealing a small compartment that held two keys and a sheet of paper. After a glance at the clerk, Starkman took the keys and paper.

The paper was a computer-printed duplicate of the one he had filled out, with an address at the top and a receipt and photoreproduction of his signature at the bottom. The address on the keys matched that on the form. The keys were rather peculiar, and Starkman studied them more closely.

Besides the ordinary teeth that worked a pin-tumbler cylinder, he saw that there were tiny plastic insets. He had no idea exactly what they were, but he already felt stupid enough about his dealings with the clerk, and did not ask. He was much more interested in getting some rest. He stuck the keys in his pocket, asked the clerk for directions, and followed them.

The sun was up and the streets were beginning to show signs of life, but no one paid any attention to him; his new clothes served him well in that respect.

The apartment was just as the clerk had promised, easy to locate, facing a courtyard, on the third floor, decorated in several shades of light brown, with phone and a small television set—or at least what Starkman considered a television set, though a little experimentation demonstrated that he could not get anything that resembled the broadcast TV he remembered, or even what he had seen at the Underground headquarters. There was no antenna, but only a cable hookup, and he could get five stations—three text and two picture: news, weather, financial reports, news with visuals and commentary, and sports. A soccer game was in progress. The current big news involved a broken irrigation pipeline in the Sahara. Weather seemed rather irrelevant under a dome.

The clerk had not mentioned that the previous occupant had left the place a mess and no one had been sent to clean it up.

A pile of old candy wrappers occupied the center of the floor, and the kitchen sink held two unwashed pans, both battered and burnt, neither one watertight. There were dark smudges on the walls, dust on the furniture, and bits of debris everywhere.

Starkman looked it over and shrugged. He could clean it up later; it wasn't all that much worse than his place near Pittsburgh had gotten at times. It seemed almost homey, save that there was no fireplace or stack of firewood. He settled on one of the dusty brown chairs and made himself comfortable.

For a few minutes he simply relaxed and let the accumulated tension and discomfort of the last few days drain away somewhat. It was the first chance he'd had to sit quietly alone since he saw the starship cruise overhead, back in Pennsylvania. He had always been either running or surrounded by people since that moment.

He had left the TV on, tuned to the soccer game; Nairobi was beating Kinshasa three to one, and Kinshasa seemed unable to do anything about it.

He had never been much of a soccer fan, even as a kid. He wondered whether he should have signed up for some of the entertainment options. It looked rather as if things were going to be boring here; he had nothing to do but sit.

No, he corrected himself, that wasn't true. Sooner or later he would have to find work; the money the Watchman had given him wouldn't last forever, particularly not if everything was as expensive as the apartment. He wondered just what the actual conversion rate between units and dollars might have been.

Kinshasa's goalie deflected a kick into the path of a charging Nairobian, who promptly booted the ball home, widening his team's lead. A few seconds later the clock ran out, for a final score of four to one.

Starkman began thinking that he might just want to catch up on his sleep. He hadn't gotten to sleep since he left the ship, and though he'd apparently gotten far enough off schedule during his captivity that he'd last awoken in what had proven to be midafternoon of the previous day, he was tired and ready for a nap.

Of course, he wouldn't want to continue sleeping during the day as a regular thing, but surely, he told himself, he could adjust later. At the very least he could take a nap.

An announcer appeared on the screen, saying, "Well, that's our game in the African League for today; stay tuned for a recap of the highlights, to be followed by championship boxing from Casablanca. First, though, we have this public-service bulletin."

After a nap, he told himself, he'd see about getting some food, and then look for work. He had no idea what work he might be qualified for; he'd had no real training in anything, and had never managed to hold a job for very long.

A different announcer appeared and said, "The search continues for the missing man, John Starkman. Authorities in Capital report that this man is believed to be carrying a mutated flu virus first discovered in the area that was formerly known as Pennsylvania, North America, where he straggled. This virus, possibly of artificial origin and intended as a bacteriological weapon, is highly contagious and has an extremely high mortality rate, so that it is imperative that the missing man be apprehended and treated immediately if an epidemic is to be avoided. Mr. Starkman is described as being approximately 180 centimeters in height, weighing between seventy-five and eighty kilos, and when last seen wore a full beard and shoulder-length brown hair, and old-fashioned clothing, including a brown winter coat, flannel shirt, and denim pants. Apparently due to a deformity of the eyes, he wears mirrored sunglasses at all times. If anyone sees this man, or has evidence pertaining to his whereabouts, please report it immediately to the local authorities. Do not approach him; he is believed to be dangerous, and the disease is, we repeat, highly contagious."

Starkman stared dismally at the screen. So much, he thought, for any chance of finding an honest job. The description would fit millions of men, and the clothes were gone, but the mirrored sunglasses were a giveaway. Any stranger with mirrored sunglasses would be under suspicion immediately unless he could prove his eyes weren't deformed.

It wasn't exactly a deformity, but Starkman could not prove his eyes were normal.

The announcer paused, then continued, "As a result of the search for John Starkman, strict limits have been put on travel to and from Capital and the areas surrounding it. These are strictly a temporary measure, and will be used only until the missing man is found. Anyone who must travel to or from Capital, for any reason, will be asked to submit to a physical examination, and all vehicles and luggage will be searched. The government apologizes for these inconveniences, but they are necessary quarantine measures."

It was a very good thing, he decided, that he had gotten out as quickly as he had.

"Once again, anyone seeing a stranger fitting the description of this man is asked to report it *immediately* to the local authorities. The other stragglers brought back with him are being located and quarantined, and it is hoped that questioning of these people will enable the authorities to provide a composite picture and more detailed description. If you are one of these people, please contact your local authorities; we ask for your full cooperation for your own health and safety. We ask all citizens to cooperate with the government during this crisis. All transmission media will be carrying periodic bulletins."

If he had ever had any serious doubt about the story the Watchman and Watcher One had told him, it was gone. It was obvious that the Rebel government of Earth was going all out to find him, even at the risk of antagonizing its human subjects. He was sure that there must be a strong undercurrent of resentment against the alien conquerors, however beneficent they might seem—not just anger at being ruled by outsiders, but the ferocious envy that must be felt in response to their manifest technological superiority. The Rebels had made every effort to appear charitable, relying on gratitude and self-interest to keep humanity in line, but Starkman knew that over time, gratitude wears thin and people resent charity. Screwing up travel arrangements, searching belongings, and running constant bulletins might get results, but it would rub people's noses in their powerlessness. The longer it went on the more they would resent it, and the more ineffective the government would appear, being thwarted by a lone man.

The bulletin was over, and a voice was trying to whip up enthusiasm over the sight of an empty boxing ring and a restless crowd. Starkman got up and switched to the news text channel.

The Sahara pipeline was repaired. The Governor had announced a plan to dome and reinhabit Houston, Texas. John Starkman was still missing, and all citizens were asked to cooperate in the search for him.

It might not be the secret of triggering supernovae, but he obviously had something the government wanted very badly, and he could think of no explanation more reasonable than the Watcher's. The creature had had no reason to lie, so far as he knew. It had behaved throughout as if it were exactly what it claimed to be, even to releasing him here in New Denver. If it had wanted him dead, it could have had him killed a dozen times over. If it had been lying, it was too subtle for him to outthink.

Therefore, at least for the sake of argument, he would assume that everything the Watcher and the Watchman had told him was the literal and exact truth. He carried in every cell of his body the secret of destroying stars.

That was a truly hideous thought. So long as he lived, the world would be in danger, and hundreds of other worlds as well. He could not find it in himself to worry very much about other planets; they were not quite real to him. The events of the last few days had been so sudden that he had not yet had time to absorb them all, and although he knew intellectually that he had been dealing with beings from other worlds, he was not yet emotionally convinced of the reality of those worlds, let alone that they deserved his compassion.

He disregarded the worlds of the Galactic Empire and concentrated on his own little planet, and on himself.

As long as he lived his world would be in danger, because once the Rebels had what they wanted it would be good military tactics to destroy him, so that he could not be used by the other side, and to destroy Earth, in case there were other yellow-eyed humans not yet located. The simplest way to destroy the Earth, he was sure, would be to blow up the sun; that would serve to test out the data, and would get any stray Terran space-

craft that might otherwise be missed. Whatever the
method was that he carried in his genes, Starkman had
little doubt that the aliens would be able to use it in
short order; besides their own fleet, the size of which
he could not judge save that it was large enough to
have easily conquered a planet, they had at their disposal all the resources of Earth and all the raw material
in the solar system.

If he were captured, and his genes really did hold
the information the Rebels wanted, not only would his
own death be certain, but the entire human race would
probably die.

If the information had been mutated away he was
unsure what would happen. They might kill him anyway, since he knew so much that they did not want
known. He was sure they could cover up a murder easily, despite their insistence on not harming humans.
They would not destroy Earth until they were absolutely certain that what they sought could not be found,
but they might well destroy him, and that was something to be avoided. His life might not have cosmic
significance, but it was important to him.

Besides, there was no way he could know whether
or not the data was there until it was too late.

His only hope was to avoid capture indefinitely.

He snorted. That was easy enough to say, but he had
no idea how to go about it. He had a few thousand units,
but very little else. He could not risk taking a job. He
should not even stay in this apartment he had just paid
too much for, he realized; the clerk, when he heard the
bulletins, might remember the sunglasses and note the
similarity of names and call someone to report the address.

He might already have done so. The Governor's flunkies might already be on their way.

Starkman pulled the flashbomb out of his pocket and
studied it, then shoved it into his shoulder bag. He was
not yet desperate enough for suicide. He began to think,
though, that when the time came, he might prefer suicide to capture after all. At least the human race would
have a little more time that way, and all he would lose
would be a few months at most.

The video screen was running another bulletin, reporting that he might have escaped from Capital before the blockade was in place, so that although the blockade would continue, citizens throughout South America were asked to look for him, and all transoceanic craft would be searched as well.

The phone rang.

Startled, he stared at it.

It rang again, and he relaxed slightly. It had to be a wrong number, he decided, or a call for the previous occupant. Nobody knew where he was. Nobody knew him at all.

Except, he recalled, for the clerk. Maybe he was calling about something—the cleaning, perhaps.

Maybe the clerk had called the police, then had second thoughts. At the third ring, Starkman picked up the receiver.

"Hello?" he said hesitantly.

"Starkman, get out of there! Fast! The rental agent's called the cops!"

It was the Watchman's voice.

"How do you know?"

"We've got Max monitoring all calls to the government, and that's the only report of you in New Denver so far, so we were pretty sure it *was* you. Who else would be renting an apartment at six in the morning wearing mirrored glasses? So we had Max trace down your new phone. Now, get *out* of there!"

"Right." He dropped the phone, picked up his shoulder pack, and ran for the door. The corridor outside was clear, but the elevator was on the ground floor; he decided he could not take the chance that it was picking up government clones, and headed for the fire stairs.

The steps were dusty, the walls bare concrete; it was obvious that when the building was finished nobody had bothered much about the stairway. He realized as he hurried down the second flight that he was leaving footprints in the dust, but he saw no alternative; he didn't dare stop to wipe them away.

At the ground floor he stopped, caught his breath, and peered cautiously out through the tiny window in

the fire door. He saw no one. He pried the door open a
centimeter or two and put an eye to the crack.

The elevator was on its way up, and he saw no sign
of anyone in the lobby. Acting as casually as he could,
he opened the door and stepped out. The lobby was
empty. He turned toward the street.

Two young men in blue denim lounged uncertainly
just outside the building's entrance; he could not see
their faces, but only the edges of their bodies showing
around the frame of the glass doors.

He had hoped that the aliens who ran the government were sufficiently ignorant of human behavior to
have not reminded the clones to post a guard at the
door, but apparently they weren't—either that, or they
considered such a precaution simply elementary tactics
applicable to any intelligent species. He was sure that
the clones wouldn't have thought of it themselves, and
took it for granted that it had been specifically ordered
by a superior.

After his escape from the old processing center it
was a safe bet that that superior would also have insisted on guards at any back door that there might be,
as well. He would have to find some other way out of
the building.

The clones were not very much to worry about. The
aliens were obviously intelligent and determined, but
perhaps they were sufficiently unfamiliar with Earth
to overlook some possibilities. He recalled that none of
the buildings that the aliens themselves ran that he
had seen had any windows; the ships, the Welcome
Center, and the old processing center had all been windowless except for small panes in the doors. If he could
reach a window he might get out unseen.

Most of the ground floor was shops that did not communicate with the lobby he was in; those were no help.
He looked around.

There were five doors opening off the lobby. The main
entrance was at the far end. He had just entered from
the fire stairs in one corner, near the elevators that
faced the entrance. A third door was the emergency
exit opposite the stairs; wherever that led, he was sure

that there would be guards posted along the way somewhere.

The other two doors led to public restrooms. They
might be windowless—but they might not, and even if
there were no windows there might be vents of some
sort. He headed for the men's room, then stopped.

He wanted to evade his pursuers completely; he knew
that they had incredibly effective means of tracking
him. Every little bit of confusion would help. He walked
calmly into the women's room.

There were four stalls, three sinks, and no windows.
A metal grille was set into the wall at the far end.

He considered using the flashbomb to blow himself
an opening, but decided not to waste the little gadget
on anything so trivial. He dug his old pocket knife out
of his pack and ran it around the edge of the grille.

There were four fastenings; a little prying demonstrated to his satisfaction that they were spring clips.
One by one he worked them loose.

He heard footsteps outside, and pried harder.

There were more footsteps, and voices, but no one
entered before he had worked the grille loose. The opening he had uncovered was big enough for him to crawl
through, and he could see that the shaft beyond was
not completely dark. He hauled himself up and into it.

The shaft was vertical; he found himself hanging in
midair, his hands on the sides of the shaft, his feet still
on the lower edge of the opening. He was facing another
ventilation grille on the other side of the shaft, identical
to the one he had removed.

He rearranged himself carefully, first moving his
feet forward into the shaft and lowering himself back
so that he was sitting in the opening he had entered
through. That blocked off most of his light, and left his
backside a perfect target should anyone enter the restroom, but it seemed necessary. Then he worked his feet
up the opposite side until they were resting against the
other grille, forcing his back up against the upper edge
of the opening. With his hands braced firmly to either
side, he kicked out as hard as he could.

The spring clips holding the other grille gave, and
it fell rattling onto the floor beyond. He fell after it,

and landed sprawling on top of it, staring at a tiled floor.

There were loud female voices shouting at him. He scrambled to his feet and realized that he was in another women's restroom.

"Sorry, folks," he said, and dashed for the door.

He emerged into a small restaurant; a few patrons glanced up at the clamor the two women who had been in the restroom with him were making.

"My mistake," he said, and walked as calmly as he could out the door onto the street.

Off to his left a pair of blue-clad clones were guarding a fire exit; he turned away quickly, before they could spot the mirrored sunglasses that had somehow stayed in place throughout his escape, and walked off up the street adjusting his shoulder bag.

Chapter Eleven

There were guards at every exit from the dome. The building he had escaped from was being searched repeatedly, room by room, though of course the two opened grilles had been discovered almost immediately; apparently the government thought he might have doubled back. The public-address system was making announcements concerning him one after another. Checkpoints were being set up all over the city; hundreds of clones had been flown in, and robots as well.

He had covered himself by browsing in bookstores and video shops, looking through department stores, blending with crowds wherever possible. He wished that he had some way of finding out what was really happening; the government bulletins were not very informative, and there was nowhere he could sit down and watch a regular news program. The news might be censored, but it was still independently run, not another branch of the government.

He knew he couldn't hide forever; in fact, he doubted he could last out the night without someplace safe to sleep. For the moment he wanted time to think, but he was too busy running and hiding to find any.

There was a small park at the edge of the dome, looking out over the surrounding mountains; it was not a particularly safe place, since clone patrols were steadily circling the dome in case he tried to cut his way out, but it was quiet and peaceful, and he paused there for a moment.

It was obvious that he needed a miracle. Unless the aliens gave up or toned down their search, they were going to find him; there was very little doubt of that. They were not going to give up; he had what they had come to Earth for. If they *did* give up, it could only be because they had found the data somewhere else, which would be even worse news than his own capture—after all, he still had the flashbomb in his pack, and could suicide for the greater good.

Of course, it wasn't likely that after a dozen years of searching, two people with yellow eyes would turn up in such quick succession. Starkman certainly hoped that there were no others.

If the aliens captured him, then either they would get the information they wanted or he would die—probably at his own hands, with the flashbomb. Neither prospect appealed to him at all.

If they got the information, then they would almost certainly destroy him and the rest of the world to keep it secret. Then they would go home and fight and probably win their civil war.

If it weren't secret, there'd be no need to destroy him, and the war would probably go on without a winner, but with a lot of losers.

It wouldn't go on very long, though; there were only so many stars to blow up in the Galactic Empire. And there would be no reason to kill him or destroy Earth.

That possibility began to run around in his thoughts.

Then, abruptly, all thought of interstellar warfare vanished; a young girl was running along beside the railing that separated the park from the service area than ran along the inside of the dome. There was something very familiar about her; for a moment he thought that his sister must have had a daughter, but then he recognized her.

"Kathy!" he called, "Kathy Saslov!"

The girl turned around and stared at him in astonishment. After a moment recognition dawned. "Mr. Starkman!" she said, "Everybody's looking for you!"

"I know. Is your mother around?"

"She's over there, with Charlie," Kathy answered, pointing back the way she had come. "You look different with your hair cut. Are you really sick, like the government says?"

"No, I'm not. I've got something the government wants, and they think that people will help them more if they say I've got some sort of horrible disease."

"Oh." She considered. "I figured it might be something like that, because if you were sick, how come we didn't catch it? They asked us a lot of questions about you at the big government building in Capital this morning, but I didn't tell them anything."

"Thanks. I don't guess there was much you could have told them, anyway; you didn't know where I was any more than they did."

"Well, yeah, but I didn't tell them anything anyway."

"Thanks. I appreciate it. Where did you say your mother was?"

"Back there." She pointed again, and Starkman followed the direction indicated, strolling along casually in case anyone was watching. He waved farewell to Kathy, and she waved in return.

The railing ran along a gentle, almost imperceptible curve in the direction Kathy had pointed, with a broad walkway alongside, and Starkman could see that there was nobody on it for quite some distance. A smaller walk branched off a few meters away, however, and curved off around a clump of trees, out of sight. He headed for that, and as he rounded the trees he saw Jenny and Charlie, walking hand in hand toward him, arguing about something. Like Kathy, they wore new, modern clothing, but were very definitely recognizable; both still wore their hair long, not in the shorter style prevalent in Capital and New Denver. Jenny's hair flowed down her back and around her shoulders in a thick black wave; aboard the ship it had been largely hidden by her old coat. He found it wonderfully attractive.

"Ms. Saslov!" he called.

She looked up, and made a wordless noise of surprise.

He continued to walk toward them; as he drew near she asked, "What are you doing here? I thought you were still in Capital!"

"I got out before they set up the blockade," he answered. "Except now I'm stuck in New Denver, and they've got this place blockaded, too."

"Oh." She looked flustered. "That must be what all those men were for when we arrived."

"Look, if you want, I'll go away and leave you alone."

"No, that's all right."

"I'm glad to hear you say that. It's good to see a familiar face."

She smiled. "I'm hardly that familiar; we only met on the ship."

"You're a lot more familiar than anyone else I've seen lately."

"I suppose I am, at that." She smiled again.

There was an awkward moment of silence, and the smile faded away.

"Why are they after you?" she asked at last. "I know you're not sick; I was with you on the ship for three days, and you didn't have any five-day flu, or you'd be dead by now."

"It's complicated. They need someone with yellow eyes."

"Why?"

"It's very complicated; I don't want to take the time to explain here."

"Where are you staying?"

"Nowhere. Would I be out in the open if I had somewhere safe to go? I rented an apartment, but they found it."

"We have an apartment. I guess you could stay with us for a while."

"I was hoping you'd say that, I'll admit. It might be dangerous, though; they'll stop at nothing to get hold of me."

She waved a hand in dismissal. "I'm not worried about that."

"I hoped you'd say that, too. Are you doing anything in particular right now?"

"I was just taking the kids for a walk; we wanted to see more of the city."

"Would you mind very much if the walk were cut short? There'll be a perimeter patrol along any minute, looking for me, and I'd like to get inside somewhere as soon as possible."

"A perimeter patrol?"

"Yes. They've got squads of zombies checking around the edge of the dome to make sure I haven't slipped out somewhere."

"My Lord, they *are* looking for you, aren't they? We're on top of a mountain! Or are there always patrols, just on general principles?"

"I don't know for sure," Starkman admitted. "It doesn't really matter; the results will be the same if they find me."

"That's true." She looked about, then called, "Kathy! Kathy, come on! We're going back to the apartment now!"

Kathy appeared, strolling around a bush, and asked, "Is Mr. Starkman coming with us?"

"Yes, he is; he hasn't got anywhere to stay, so we'll be putting him up tonight. Now, come on."

The girl came, as did her brother, who had yet to say a word in Starkman's presence. Jenny led the way, with Starkman at her side and the children a pace or two behind, out of the park and into the city.

The sun was reddening in the west and the streets were awash in the shadows of the buildings, and swarming with ground cars as the working population headed for home. The sidewalks were full, and Starkman had little trouble in going unnoticed in the crowds. It helped that he had a woman and children with him, and even more that he wore modern clothing. He knew he would never have avoided capture even this long without the aid of others.

There were no aircars in sight; he wondered whether they were legal under the dome. They certainly weren't necessary; New Denver was not as big as its namesake. Jenny had just pointed out a tower ahead as her new home when the public address speakers crackled to life.

"Attention, please. This is an official announcement. To aid in the ongoing search for the missing plague-carrier, John Starkman, the government has decided to place a temporary ban on the wearing of sunglasses. Until further notice, wearing sunglasses will be grounds for detention and a fine of up to ten thousand units, effective immediately. This directive includes all devices that serve to conceal the eyes, or to modify their color. All citizens are asked to cooperate."

Starkman glanced around. People were stopping and staring at the speakers; the announcement must seem utterly senseless to the average citizen, he was sure. Still, people had a tendency to obey authority automatically, at least until it became actively inconvenient or uncomfortable.

"Come on," he said, "Let's get inside." He picked up his pace, hoping that the children would be able to keep up.

After a brief hesitation, the announcer was continuing. "The missing man, John Starkman, is described as being approximately 180 centimeters in height, weighing around seventy-five or eighty kilos, and when last seen was wearing a full beard and shoulder-length hair, all brown. His outstanding peculiarity is a hereditary discoloration of the eyes, which are yellow. Any citizen seeing any person with yellow eyes is asked to report the sighting immediately to the nearest government office. A priority code has been established in the comnet; if you see this man, and no government official is convenient, tell Max. Max has been programmed to react appropriately to the name Starkman; we regret any inconvenience this may cause to other persons with similar names. John Starkman, recently brought in from Pennsylvania, is believed to be carrying a mutated disease virus. He was last seen, confirmed, near the Capital spaceport, but reliable unconfirmed sightings have been made in the cities of Capital, New Denver, and Rio de Janeiro. Citizens of these cities are asked to be vigilant."

"Rio?" Jenny glanced at him. "Were you in Rio?"

"No; that one's a mistake, I guess." It occurred to him that the Watch or the Underground might be faking sightings to confuse the government.

They reached the door of the building and hurried in; Jenny led the way to the elevators.

They managed to get a car to themselves; when the door had closed, Jenny said, "We're on the top floor. The kids wanted the view." She punched the appropriate button, and the elevator rushed upward.

Starkman was less unready this time, but the sudden vertical acceleration still bothered his stomach slightly.

The Saslov apartment had a large living-room/dining-room/kitchen, three small bedrooms, and a single bathroom. One end of the living room was mostly glass, providing a splendid view of the city from forty stories up—though there were several taller buildings blocking out portions of the landscape.

There was a computer console similar to that Starkman had seen in the Watchman's apartment; Jenny noticed him studying it, and remarked, "I can't use it very much myself, but I thought it would be good for the kids to have; it's got as much service as I thought I could afford. I haven't really gotten used to this new money, though, and I haven't gotten work; the government made me a loan."

"I have some money; I can pay for my stay here."

"It would be welcome, I admit it. Umm... Where did you get any money? The government wouldn't have loaned you anything."

"No, they didn't, but a friend in Capital did."

"Oh." She thought that over for a moment. "If you have friends in Capital, what are you doing here?"

"Hiding. They were already searching Capital. Now that they're searching here, too, it doesn't matter very much, but at the time coming here seemed like a good idea."

"You still haven't told me why they're looking for you. Why are you so important? *Are* you a carrier of some sort?"

"Not exactly." He settled onto a convenient couch. "Have a seat, and I'll explain."

She sat, and he explained. She asked questions along the way, and the sun was down well before he had finished. Kathy and Charlie listened along with their mother.

After he had described his escape from his own apartment and a little of his subsequent wanderings, he wound up with, "And then I saw Kathy in the park."

Jenny sat silently for a moment, they said, "Do you know how paranoid all that sounds?"

"I guess so."

"Are you sure that's the truth?"

"No. I can't be sure. But if it's not, then why are they looking for me? Maybe I am a little paranoid, but they really *are* out to get me, and you know it."

"It's all so bizarre. The world conquered by aliens looking for some secret formula planted in your chromosomes."

"I know. If the Watcher told the truth, though, that is what happened—and I can't think of any better reason for the aliens to have come here. It explains everything, doesn't it? The sudden ice age, the convenient arrival of the fleet, the blood samples, everything."

"And if they find you, and get what they're after, you think they'll destroy the entire world."

"Yes."

"Do you know what I should do? I should kill you myself. I don't want my kids or myself to wind up flash-fried by an exploding sun, and you say that's what will happen if they find you alive."

"Look, if they capture me, I promise I'll kill myself." He fished out the flashbomb. "I do have this."

"That's fine if you get a chance to use it."

"If that's the way you feel, maybe I had better just go, before you work up the nerve to commit murder."

"No, stay. It's all right, really. I'll hide you as long as I can."

"Thank you."

"I don't know how long it will be, though."

"I realize that. I'll be grateful for whatever you can manage. Maybe I can think of some way out of this mess."

"I hope so." She glanced uneasily at the flashbomb. "I have to get the kids some dinner; you just make yourself at home."

"Thanks."

She rose and headed for the kitchenette; Starkman leaned back and rested.

He had intended to just rest his eyes for a few seconds, but the next thing he knew he was being awakened by Charlie tugging at his hand.

"Dinner's ready," the boy said.

Jenny was silent throughout the meal, but Kathy

and Charlie kept him busy answering questions. He told them of his escape from his apartment in dramatic and sometimes exaggerated detail, and elaborately described the alien he had seen.

He found himself enjoying it. It had been a long, long time since he had been able to hold a friendly, relaxed conversation with anybody, and never in his adult life had he had a chance to speak at any length with children. The few children he had encountered during the years of collapse were not permitted near him; the later ones had learned to be wary of any other human in the wilderness, and the earlier ones were kept away by their parents, who seemed to fear that Starkman's peculiarity might be contagious.

He wished it *were* contagious; then he wouldn't have been the only one.

After dinner, the children played with the video, calling up miscellaneous data, flicking from channel to channel and giggling at the more absurd images that appeared. Jenny explained that it was only the second time they had ever been able to use one; the first had been a half hour she allowed while she was out buying groceries immediately after they had rented the apartment.

More or less by accident, Kathy tuned in a newscast, and Starkman heard his name mentioned. He was suddenly interested.

"...failure to apprehend this dangerous fugitive, the Governor has decided to provide additional incentive. Any person providing information leading to the capture of John Starkman, alive and uninjured, will receive a reward of one million units in cash. In the event of multiple reports, duplicate rewards will be given to all parties involved."

Kathy turned and stared at Starkman. "Mother? That's a lot of money, isn't it?"

"Yes, it is," Jenny answered. "But it's probably a lie again, like when they said he was sick. Besides, Mr. Starkman is our friend, and also, if he's captured they may kill all of us."

"I know, but what if somebody saw him come here?"

"We'll just have to hope that no one did."

Kathy nodded, and changed the channel as the announcer repeated Starkman's description.

They finally settled on watching an old detective movie, but missed the title. The children required endless explanations of how things had worked before the cold came. They did not know what a detective was, or for that matter any policemen, and had a great deal of trouble in thinking of anyone who worked for the government as a "good guy"; Jenny explained to Starkman that there had been a series of dictators in their area before the cold got impossible, and although the last was gone before Charlie had been born, she and her husband had consistently spoken ill of governments as a result.

When the movie had ended, and Jenny had explained what a prison was and that policemen didn't just kill murderers, the children were put to bed while Starkman flicked through a few of the video channels on his own. The offer of a reward was repeated, and the stricture against sunglasses; his description was given on no less than four of the hundred-odd channels as he rushed through. One had a bad composite sketch of him as he had appeared before changing clothes and trimming his hair and beard.

In the background he was aware of the children marveling at water that didn't have to be pumped or heated, and toothpaste that didn't have to be thawed out. "It's all warm and gooey!" Charlie exclaimed at one point.

Finally both were tucked away behind closed doors, and Jenny returned to the living room.

"Where do I sleep?" Starkman asked quietly. "Here?" He gestured at the couch he sat on.

"Not unless you want to."

"Where, then?"

"It's been a year since my husband died, John."

"I've been alone for a decade; I'd gotten used to it." He stood and reached out uncertainly. "That doesn't mean I want to be alone any longer."

"You don't have to be." She moved forward into his arms. He kissed her awkwardly, then she led the way into her bedroom.

Chapter Twelve

Starkman awoke the next morning with a feeling of unreality; the warmth of a woman's body in the bed beside him seemed more like one of his better dreams than like the waking world.

The night before had not been a complete success; he had never had much experience with women, and had been alone for ten years, so he had been somewhat clumsy and inept. Still, he looked back on it with pleasure. Though he had not wanted to leave Pennsylvania, and was certainly not enjoying his current status as a fugitive, there were definitely redeeming features to his return to human society.

Trying not to wake Jenny, he slid out of the bed and found his clothes. He had a very limited wardrobe; he had not wanted to weigh himself down while traveling, but had planned on buying more once he had found himself a place to stay. He had underestimated the intensity of the search for him, however, and now he did not dare step outside the apartment. That left him with one pair of slacks and two shirts given him by the Watchman, and his own sunglasses, boots, socks, and underwear. Having little choice, therefore, he put back on everything he had worn the day before except for the lavender shirt. His second shirt was pistachio green, a color that did not appeal to him; he marveled briefly at the aberrations of fashion.

Dressed, he wandered into the kitchen area of the main room and began looking through cupboards for something suitable for breakfast.

There wasn't much; the Saslovs had been living there

less than a day, hardly long enough to accumulate much of a stock. He found a frying pan and half a dozen eggs, and fried himself two.

As he slid them onto a plate Charlie appeared in the doorway, and a third egg went into the pan. Kathy turned up a few seconds later and got the fourth. The last two went back into the refrigerator, as Jenny showed no signs of stirring.

The kids were generally quiet, but Charlie did ask whether Starkman would be staying with them forever.

"I don't know," he replied. "But I don't think so."

When the frying pan had been rinsed out and put in the dishwasher—a convenience Kathy pointed out; Starkman would not have noticed its presence himself—he settled on a couch and turned on the video, tuned to one of the independent all-news, image-and-voice channels.

He caught the tail end of a story about one of the domed cities up north, and listend with real interest to an announcement concerning the four captured revolutionaries who had shot it out with government troops the night before last. The Governor had decided to drop all charges and pardon them, since no one had been seriously hurt.

Starkman wondered at that; why were there only four? Who had gotten away, or else, who was the government secretly holding? How could the government claim that no one had been hurt, when the television reports he had seen live on the spot showed several clones lying in pools of blood? Did the government not consider clones to be worth worrying about?

The announcer on this particular station had obviously asked himself that last question or two. He spoke over a videotape of the firefight, showing close-ups of "Imperials" looking very much like corpses, asking aloud what the Governor considered serious injury.

Letters appeared on the screen, reading, "The government has not approved of this report, and does not consider it accurate or in the public interest."

The announcer finished his remarks, and continued, "Today's editorial involves a related story; the government is still searching for the missing man calling him-

self John Starkman, who was believed to be involved
in this recent shooting incident in Capital. This man
is reportedly infected with a mutated flu virus, and
extremely dangerous, yet has somehow eluded capture,
and no other cases of the disease have been reported.
In its continuing effort to locate him, the government
has now banned all intercontinental travel, and all in-
tercity travel in South America. A reward of one million
units has been offered for information leading to his
apprehension, and restrictions on sunglasses and concealing garments have been imposed worldwide. House-
to-house searches are now under way in Capital, Rio
de Janeiro, Brasilia, and New Denver. It is the editorial
opinion of this station that the government is overre-
acting to a possible threat to the public health; the
measures taken seriously interfere with the rights of
many millions of innocent citizens. The representatives
of the Galactic Empire have heretofore carefully re-
spected human rights; why has what should have been
a minor incident been blown all out of proportion? Can
all this fuss really be about one sick man, who has not
yet infected a single other person? We call on the Governor, and whatever officials have introduced these dra-
conian measures, to reconsider their actions before they
seriously alienate their constituents, and thereby ham-
per their future efforts on behalf of the world's people."

Starkman turned off the set.

He had not realized that a house-by-house search
was being made; that would almost certainly locate him
before very much longer.

It was obvious that he was not going to be permitted
to hide. So long as the Rebel-run government of Earth
remained in power, its every resource would be dedicated to finding him. If that went on, sooner or later
they would find him; a lone man simply could not pre-
vail against such odds.

There was no way he could overthrow the government quickly enough to save himself, either, so far as
he was aware—and even if the government were to
fall, so long as the Governor and his fellow Rebels were
anywhere on Earth they would pursue him. The fact
was, he knew, that they didn't much care about who

ran this world, save that being in control of the government improved their chances of capturing him—him, John Starkman.

That meant that sooner or later they were going to get him, unless he were to die first. Even then, his death alone would not be sufficient. His body would have to be destroyed as well.

If that happened, the government would have no proof that he was dead, and would presumably go on searching indefinitely—or at least until they managed to anger the populace of Earth enough to trigger a rebellion. That might be all very well for humanity, but it wouldn't do him any good. Suicide simply did not appeal to him.

So they were going to get him, and the Rebels would have the secret of the supernova bomb.

He had thought about this the preceding day, in the park by the dome wall. If it weren't a secret, then there would be no problem.

"Good morning," Jenny said from the door of the bedroom.

"Hello," he answered, "We left you a couple of eggs. Do you know anything about the postal service around here?" A half-formed idea was wandering about in his thoughts.

"No. Why?"

"I think I may want to mail a package, and I wondered how long it would take to get there."

She shrugged as she headed for the kitchenette. "I haven't the slightest idea."

That idea didn't seem very promising. "Do you know any doctors, perhaps?"

"No. We just got here, John—just a few hours before you did."

"That's true." He looked at the computer keyboard and video display. Somewhere in there, he was sure, would be all the information he might need to solve his problem.

He had been thinking of taking samples of his blood and mailing them to whoever he could find who might be able to analyze the DNA and send the information out into space, for eventual receipt by the real government of the Galactic Empire. Upon consideration, though, he realized that it almost certainly wouldn't work. The mails might well be searched; furthermore, he knew nothing about how long it took chromosomes to break down and decay, or how blood samples would have to be packed to preserve the genetic information in the white cells. He did know that red cells did not have normal nuclei, and were therefore probably worthless for his purposes.

And, he realized, he knew no one who would be able to transmit the date except, perhaps, Watcher One.

And no ordinary doctor would have the equipment for genetic analysis, so he couldn't just have the job done himself and mail the data.

The basic concept was still good, he was sure; if he could transmit the data to the Empire, then there would be no harm done if he were to be captured by the Rebels. The balance of power—or perhaps of terror—would be preserved. The secret would no longer be a secret, and the Governor would have no reason to kill him or to harm Earth—and for that matter, no reason to stay on Earth. He hoped that, if he did manage it somehow, the Governor wasn't sufficiently human to kill him out of spite.

He had no way of analyzing his DNA, or of transmitting the information, but the Watch probably did. He would have to contact them and suggest it.

They would, he knew, probably dislike the idea. After all, they had been doing everything they could to suppress new weapons technology, and he was going to suggest that they give it away instead.

He saw no other way to preserve his own life, however. And since he didn't want to commit suicide, it might well be the only way to preserve the lives of everyone else on Earth, as well. If he could convince the Watcher of that, it would probably aid him, albeit reluctantly.

He couldn't be absolutely certain, however; he wanted as much back-up for his plan as he could provide. He wished he knew of an Imperial spy on Earth. He wondered if there were any.

He also wondered how long he had before the door-to-door search reached him. He reminded himself that he had no time to waste.

He needed more information.

The video, the phone, and the computer were all part of Max, the comnet, and he had a rough idea of how to use its more basic functions. He also knew the code to reach the Watchman—what would once have been his telephone number. Jenny's service did not include a voder, unless there were one built into the phone, so that communication with Max itself or with any of the library or teletext functions would have to be done with the keyboard and display screen, rather than by spoken word. Starkman had no idea whether he might have used the phone, or time to figure it out. He wished he knew how to type.

He pecked out a command for library access in the form the Watchman had shown him, and then a request for information on genetic analysis; to his dismay, the screen was immediately crammed with tiny type, listing hundreds of subheadings.

That was no good. He punched the CLEAR button, and stared for a moment at the blank screen. He needed help; he knew that. He would need someone who could take blood samples and analyze them. He could think of no one who might be able to help except Watcher One; genetic analysis was not something that could be done in the corner drugstore, or by the average family physician.

With the possibility of being captured at any minute, he didn't really have much of a choice any longer; he punched the PHONE button, lifted the receiver from beside the keyboard, and typed in the Watchman's code.

A voice at his ear answered, "Yes?"

"This is John Sta—Stanford," he said, remembering at the last minute that Max had been programmed to note any mention of the name "Starkman."

"Ah, of course. I remember you, Mr. Stanford."

"It appears that I am about to go to work for the government, whether I want to or not. You are aware, I assume, of the house-to-house search currently under way in New Denver."

"Yes, of course; we were worried about you."

"They haven't found me yet, but I don't suppose it will be much longer; I've run out of ideas on how to hide."

"I'm afraid we can't be much help; I don't know of any way that we could get you out of the dome. It appears that our choice of New Denver may have been a mistake right from the beginning."

"I think it was."

"We hadn't realized that they would be quite so ruthlessly efficient."

"Obviously. Though if they were really efficient they'd have gotten me by now."

"Do you plan to use the device I gave you?"

"That depends. I have an alternate plan." He went on to explain his reasoning, and concluded, "Is there any way you can help?"

"We can do the analysis and transmission, but I'm not sure that we want to. I'll have to confer with the Watcher."

"I figured that you would. Let me say, though, that I will *not* suicide if you refuse to help me. If they get me before you have a chance to act, I'll try to use your little toy; at least, I think I will. But if you people are too damned moral to help me, then you can forget about my doing any favors for you or anybody else. After all, I wind up just as dead either way, if you don't help."

There was a long pause before the Watchman answered conversationally, "That sounds an awful lot like blackmail."

"I suppose it is, but *I* never said I was moral, and my life is on the line here."

"Yours and millions of others, Mr. Stanford."

"So I'm selfish. Look, if they get me, I'll try and leave something you can use with the people in this apartment. I assume you've traced the call."

"Yes, of course."

"Good. These are good people here; you do what you can for them if there's any trouble."

"We will. I'm going to hang up now, and talk with the Watcher; we'll try and get someone out there for samples as soon as we can."

"Good. Oh, wait; one other thing. Is there any way you can delay the search, or disrupt it?"

"We're trying. We do have a few people involved."

"It's too bad you don't have a few assassins somewhere. Everything would be so much simpler if the Watch had just killed off the Rebel leaders a few thousand years ago."

"Yes, I know, but then would the Watch be any better than the Rebels are? If someone had killed the Governor the Rebels would probably never have reached Earth, but they might have discovered that the Watch existed, and by now the Empire might be in the middle of a three-way war instead of two."

"Was the Governor as important as all that?"

"Oh, yes; this whole expedition is his personal enterprise. He's the only member of the Rebel oligarchy involved. We're wasting time, though; I'll talk to the Watcher, and see if he'll permit the data to be transmitted. If he agrees, we should have someone there in half an hour or so."

"All right." Starkman hung up the phone, and punched the CLEAR button just to be sure. He glanced over at Jenny and the children, gathered at the table while Jenny finished off the eggs. She smiled at him, and he smiled back.

"Anything I can help with?" she called.

"No, at least not yet. Thanks anyway," he replied.

She nodded, and went on eating.

He still wanted some sort of back-up, in case the Watcher decided against transmitting, or the messenger didn't get through. There was nobody else he knew that he could trust at all.

Even someone he couldn't trust might be better than nothing, however, and there was one other group that he knew of that opposed the Governor. They might not have the Watch's resources, but they might manage something. He picked up the receiver again, then typed in the code for directory assistance. A voice spoke in his ear; he could not decide if it was human or computer-generated, and it didn't much matter.

"I'm trying to reach a Dr. Janet Curtis, in Capital; she works at the Welcome Center there."

"One moment." There was a flurry of electronic click-ing, and then a string of numbers. He punched them in as they were given.

There was the sound of the phone on the other end ringing, and then Curtis's voice said, "Hello?"

"Don't say my name," Starkman told her. "I'm going to assume that you recognize my voice and that you know who I am and why you mustn't say my name. I call myself Stanford. Do you understand all that?"

"What...I think so." Her tone was cautious and slightly uncertain.

"I hope so. Before I go any further, were you one of the four revolutionaries they captured and released?"

"No. That was Emilio and Raoul and Ramón and Paulo. Cheryl and Jan got out the back, and I told them that I'd been kidnapped, the same as you. They believed me and let me go."

"Good. Do you think they're suspicious of you?"

"No. Those idiots? They trust me."

"Good. I'm going to tell you a little story, and I want you to tell everyone you think might be interested. All right?"

"All right."

"Good." He told her briefly what had happened to him, from the firefight at the processing center up through the explanation the Watcher and Watchman had given him, on up to the point where the Watchman agreed to help him hide. He carefully altered every mention of the Watch, referring instead to Imperial spies. He did not see any point in blowing the Watch's cover; the Underground might not take kindly to the presence of a rival organization of that sort. He avoided detailed descriptions of people or places. He did not tell her that he was in New Denver, or anything else that he thought might help her locate him; he was afraid that if he did, the Underground would try to kidnap him again to keep him away from the Rebels.

He did explain exactly what the supposed Imperial spies had told him, and hoped that it would not occur to her to wonder why they hadn't held on to him or killed him. He made no mention of the flashbomb or suicide.

When he had finished there was a moment of silence; then Curtis demanded, "Is that all true? Is that why they're here?"

"So far as I know, yes, it's all true. They showed me those tapes. Of course, I can't be sure that the alien and his flunky didn't fake the whole thing somehow, but for myself, I believe them completely."

"Damn. I knew you were important, particularly after all the fuss that's going on right now, but I wouldn't have guessed that they came here and took over just to get *you*."

"It came as a surprise to me, too," he remarked sarcastically.

"So why are you calling? Do you want to join our organization?"

"No. You've been compromised, after all, even if they did let the four of them go. I wasn't even sure you were free; I thought that maybe they were lying, and were still holding some of you. For all I know someone's listening in on all your phone conversations, and they'll come and get me if I stay on the line too long and let them trace the call. But I had an idea that I had to tell you, an idea of how I could end this and stop running and hiding."

He explained his idea of sending the analysis of his DNA to the Empire, and went on, "The problem is that I don't entirely trust those spies. They might not be what they say they are. I want your group to help. I'm going to try and get samples of my blood to as many different factions as I can, and hope that somebody manages to get the information to the Empire before the Rebels stop them. I've already contacted the Imperial spies; now I'm asking you people. Then I'm going to start trying to find any other underground groups that might exist. Will you help?"

"I don't know. I don't like it, giving a secret weapon to another bunch of aliens."

"What else can we do?"

"I don't know. Fight them ourselves."

"We can't use a supernova weapon; the only star we can reach would fry us, too. We just haven't got the capability of fighting them." He remembered the

Watchman's assurance that as much alien science and technology as possible was being disseminated to humans, and added, "At least, not yet."

"I don't like it."

"That's too damn bad. Look, I have to get off the line; someone will call you back and make the arrangements if you agree." He hung up.

"Who's going to call them?" Jenny asked as he turned away from the console. She had finished her breakfast and was crossing the room toward him.

"You are, I hope," he told her. "You heard how I got hold of her; you can do the same. Dr. Janet Curtis, in Capital, works at the Welcome Center. You met her; she was the doctor who examined you and the kids."

"Oh, her?"

"Yes, her."

"Do you really think her phone might have been tapped?"

"I don't know." He looked back at the computer. "Maybe I can check."

He tapped the sequence of keys that the Watchman had told him would connect him to the central computer itself, then typed in, MAX, WAS THE LAST PHONE CALL FROM THIS CONSOLE MONITORED?

His message appeared on the screen, and immediately below it was the single word YES.

"Oh, damn, they'll probably be here any minute then!" He whirled away from the keyboard and stood up. "Jenny, have you got a sealable jar and a sharp knife?"

"Yes, I think so; why?"

"I'm going to take a blood sample and leave it in your refrigerator. I don't have time to find a doctor; they might be here any minute. You give the blood to the Watchman or the Underground or whoever you can."

"Oh," she said. Starkman ignored the revulsion on her face and ran for the kitchen.

He had no time for caution, and cut deeper than he had intended. He was trying to wrap a bandage around his bleeding arm when the door of the apartment was smashed in. The jar he had filled was already safely in the refrigerator.

Chapter Thirteen

"What happened?" the doctor asked as he wrapped gauze around Starkman's arm.

Starkman shrugged. "I slipped with a knife. It was an accident."

"It looks almost like you were trying to slash your wrists."

"The hell it does! I could do a better job than that; you want to cut the vein lengthwise, not across or at an angle like that. Besides, it's supposed to hurt less if you do it in warm water; if I were going to slash my wrists, I'd have done it in a hot bath."

"All right, all right, calm down; it was just a casual remark."

"I'm sorry; I'm a bit tense. I don't want them thinking I'm suicidal and posting a death watch."

"I understand." The doctor smiled; he was a gray-haired old man with a faint accent to his English that Starkman could not place exactly. "I have to admit that it made taking a blood sample pretty easy, having a messy wound like that still open and bleeding."

"I wanted to bandage it, but they wouldn't wait. I barely had a chance to pick up my bag. I don't know what the big hurry was. They've been looking for me for a couple of days now; how much difference could five minutes make?"

The doctor shrugged, and smiled again. "It's a good thing they brought you right to me; that cut might have been serious if it weren't properly attended to."

"I tried to tell them that, but they wouldn't listen. Damn stupid zombies." The doctor's smile blurred, and

Starkman realized that his speech was slightly slurred. He felt very tired suddenly, and weak.

"You just rest, Mr. Starkman. They want some tissue samples, as well, and I think they may want to question you, but right now you've lost a good bit of blood, and I think it would be best if you just rested for a while. I'm going to tell them that." The doctor rose; Starkman noticed how short the man was. He lay back on the cot and shut his eyes as the door closed behind the departing physician.

His arm did not really hurt, any more than it had hurt when he first cut it and filled Jenny's storage jar with almost half a liter of blood. The doctor had put some sort of salve on it before bandaging it.

On the other hand, he remembered very little of his hurried trip from the Saslovs' apartment in New Denver to this holding cell in Capital other than the surging waves of pain caused by the idiot clone who had grabbed his still-bleeding arm and used it to drag him.

Starkman had to admit that it had been effective; he had been in too much agony to put up any resistance at all. He had been unable even to think.

Now he felt a bit better; there was no real pain, just a dull throbbing awareness that all was not well with his left wrist, and a thick, heavy weariness that he knew was from loss of blood. He was able to lie back and consider his situation.

He might well be doomed, and the rest of the human race with him. He had no assurance that either the Watch or the Underground would get any of the blood he had spilled, or that they would be willing to analyze it and transmit the data. He was quite certain that if neither one did, or if they neglected to inform the Rebel government, then the Rebels would test out their new toy on Earth's sun as they left. He himself probably would not live even that long; once they had the information they were after he would be a liability.

He wondered if he would ever know what was to become of him, or of his world.

The doctor returned perhaps twenty minutes later and took scrapings from Starkman's right arm and mouth. He studied those peculiar yellow eyes that had marked Starkman throughout his life, checked the bandage, and tried to be soothing. Starkman, having more or less resigned himself to his fate, allowed himself to be soothed.

The doctor left again, and Starkman fell asleep.

He had no idea how long he slept; when he awoke he still felt weak, but was much more concerned about being ravenously hungry. His arm was not bothering him, which he considered a good sign.

He looked around his cell. It was nearly a cube, three meters on a side—or perhaps the ceiling wasn't quite that high; it was painted white while the walls were beige, which made it appear higher than it actually was. There was no window; light came from a single panel overhead. The door was some solid substance painted to match the walls, and with a small window at eye level, reinforced with hexagonally gridded wire. The furniture consisted of the narrow metal cot and a straight-backed chair the doctor had sat upon. There was no sign of sanitary facilities—and of more immediate interest to Starkman, there was no sign of food or water.

He started to get to his feet, then stopped and fell back to sit on the edge of the bed while a wave of dizziness spent itself.

As the moment of vertigo passed and he prepared again to rise, the doorknob turned and the door swung open.

"Hello, Mr. Starkman," the blue-clad young man said cheerily. "I saw you were awake, and thought I'd come see if there was anything I could do for you."

Starkman ignored the implied presence of a hidden camera monitoring his cell; he had half expected that there would be one. "I could use something to eat," he said.

"Sure, Mr. Starkman, you wait right here, and I'll be right back." The clone stepped back, closed the door, and was gone.

Five minutes later, after Starkman had paced out the dimensions of the little room and decided that three meters square was just about right, and discovered that he could reach the ceiling easily, so that it could not be over two and a half meters high, the door opened again and the clone entered, balancing a wonderfully full tray.

Starkman made no pretense of politeness; he was hungry, and he was being held against his will, and saw no need for good manners. Besides, he didn't like clones. He stuffed himself with roast beef, peach cobbler, and the contents of an immense pitcher of orange juice, ignoring the young man who stood quietly watching.

When he felt a bit more like himself he sat back and looked at the "alien"; the old quote, "The condemned man ate a hearty meal," ran through his head, and he wondered whether this would really be his last supper.

At his request, the clone escorted him to a tiny washroom across the hallway, and then back. As he settled down on the cot once again, the man asked, "Anything else, Mr. Starkman?"

"That depends. Where am I?"

"You're in a special section of the Governor's headquarters in Capital."

"Why am I here? Am I a prisoner?"

The clone shrugged. "I dunno, really. I guess so; they told me not to let you escape, so I guess you're a prisoner."

"Am I allowed to make a phone call? It's traditional to allow a prisoner one call." He wanted to check with Jenny and see if his plan had been carried out.

"You mean com somebody? I don't know, Mr. Starkman; they didn't say. They said that I was to make sure you were comfortable, but not to let you out."

"Well, I won't be comfortable unless I can check with my friends and make sure they're all right."

The young man looked uncertain.

"Is there any way you could get a phone in here?"

"I don't think so."

"Is there someone you can check with?"

"Well, yeah, I guess so."

"Why don't you go ask, then?"

"Okay." He stepped out, leaned back around the door to say, "Be right back," and then was gone, leaving Starkman locked in and alone again.

He sat on the cot and stared at the walls.

It seemed rather nasty to have made the cell so utterly plain; there was nothing in it to relieve the boredom of waiting. He spent some time trying to guess where the monitor camera was, and finally decided that it must be concealed behind the light panel, where he couldn't see it; he certainly found no trace of its presence elsewhere.

He tried staring through the little window, but the corridor outside was just as blank as his cell, with beige walls and white ceiling and the blank beige door of the washroom, and nothing else in sight.

The floor was tile of some sort; he counted tiles and found that there were a hundred and forty-four of them, all identical squares of off-white.

Someone had removed his boots some time while he was not fully aware; they were stuffed under the cot, along with his shoulder pack. He hauled them out, feeling slightly faint as he bent over to reach for them. He pulled the boots back on, then began checking the contents of the bag.

The flashbomb was still there, but there was no sign of his old pocket knife or the rattail comb he had bad. Someone had apparently searched it and decided that the lighter was harmless. He suspected the searcher had been a clone; who else would have taken the comb and left the bomb?

Or perhaps the searcher had tried the "lighter" and found that it didn't work.

He was still rummaging through his meager belongings, seeing what remained and what was gone, when the door finally opened again. He stood up, expecting either the friendly clone or the elderly doctor.

It was neither; instead he faced a squad of six particularly vicious-looking men. Although he could not be absolutely certain, since most of them were bearded and of about the right age, he did not think any of them were clones; they had too much character in their faces, all of it unpleasant, and were too varied in their appearance. They did wear the blue denim uniforms, but with unfamiliar shoulder-patches.

"Come on, buddy," one of them said. "Some people want to talk to you."

Starkman was in no position to argue, and he knew it. He slipped his bag onto his shoulder, expecting one of the party to object to its presence. None did.

It was too late for him to suicide, he realized; they already had his blood. Otherwise, this might have been the ideal moment, as he might never have another chance, and the six thugs were clustered so tightly about him as they moved down the corridor that he could easily have included them all in the flashbomb's explosion.

He was escorted through a series of three guarded security doors; that explained to his satisfaction why half a dozen human toughs were assigned to him now, where a single clone had been adequate to deal with him in his cell, or even in the corridor and washroom outside. Even if he had jumped the poor zombie and made a run for it, he would not have gotten past these doors.

Once out of the detention area he was taken through a maze of corridors, elevators, doorways, and antechambers. The beige walls gave way to white, and signs indicating directions to various numbered rooms began to appear at the intersections of hallways. He glimpsed other people occasionally, always in the process of scurrying out of the way of the advancing phalanx that surrounded him.

Finally he was brought into a large room that he took for a laboratory of some kind; three or four people in white coats stood by various machines, watching him enter. He reached instinctively to adjust his sunglasses, but they were not there; he could not remember whether he had left them in the Saslovs' apartment or lost them somewhere since.

Two of his escort kept a firm but not uncomfortable grip on his shoulders and arms; the other four stepped back out of the way, and he found himself facing the elderly doctor and three video cameras.

"Mr. Starkman," the doctor said with a polite nod.

"Hello, doctor," he replied.

"There seems to be some difficulty involving the tissue and blood samples we took from you."

"There is?"

"Yes, there is."

"What sort of a problem?"

"I am afraid I cannot tell you that. However, my superiors insist that these samples are not actually your own flesh and blood."

"That's silly; you took them yourself, with a camera watching the whole thing."

"Yes, I know. My superiors claim that you somehow managed to substitute someone else's tissue for your own."

"That's ridiculous! How could I do that?"

"I have no idea; but then, I have no idea how the Galactic Empire can exist, either. I am no longer ready to say that it's flatly impossible. Therefore, we are going to take some more samples, here and now, with several witnesses, both human and mechanical, and in such a way that there can be no doubt at all that the tissue is indeed your own. That means that the tissue must come from a part of you that cannot possibly have been transplanted or transfused or somehow altered." He gestured toward an operating table. "Lie down."

Starkman resisted, but the six guards had little trouble in placing him on the table and holding him down.

The doctor, closely watched by three cameras and three other people, then proceeded to use biopsy needles to draw cells from various portions of Starkman's body, from his head to his groin. The operation was virtually painless; the needles, the doctor explained, had built-in anesthetizing devices.

When the samples had been taken they were moved, under careful multiplex observation, to a bank of machinery, and placed into prepared containers. After that, they were out of Starkman's sight.

He was permitted to sit up on the table's edge and pull his shirt back down. He looked up at the cameras and asked, "Who's watching those? Can they be trusted?"

The doctor glanced up at them. "Oh, I think so; one of them is going directly to the Governor himself, they tell me."

"Oh." Starkman waved. "Hello, Governor, you alien monster, you."

The doctor cast him a quick look of surprise. "Why do you say that?"

"Why not? That's what he is, I understand. Or maybe I should say what *it* is."

"How do you know? No one's ever seen him."

"No one has ever seen him on Earth, maybe, but I've spoken to some of the other aliens."

"The Imperials?"

Starkman decided not to answer that. Instead he addressed the cameras again.

"Hey, monster, it won't do you any good to analyze my DNA; the secret is already out. I gave samples to an Imperial spy I met. Yes, I met a spy; how do you think I managed to hide for so long? Your whole fleet is full of traitors working for the *real* Empire. By now the data you're after have been transmitted; the signals must be halfway to Jupiter, and there's no way you can ever catch them. If one side is going to have a super-weapon, I figure we're all better off if you *both* do. Now, for every star of theirs you blow up, they can blow up one of yours. Or maybe you can negotiate and not blow up *any* stars, not even ours here on Earth."

The four doctors and even the guards were staring at him.

"What are you talking about?" asked a white-coated young woman.

"That's what they're after, you know," Starkman told the humans casually. "That's what they came to Earth for. Thousands of years ago one of their scientists pro-grammed a secret weapon into the genes of a yellow-eyed human. They conquered Earth just to get me. They caused the ice age just to have an excuse for taking over and hunting me down."

"That's crazy!" said one of the guards.

"Yes, I know, but what else makes any sense? Why have they pursued me so relentlessly? You all heard their bulletins and announcements. What are they doing with these tissue samples? These aliens are in the mid-dle of an interstellar war, and they came here looking for a way to win it. Fortunately for us all, the other side, the *real* 'Galactic Empire,' found out about it and planted a few dozen spies in their fleet. They've recruited hundreds of spies here on Earth. One of you is probably an Imperial spy; how else do you explain the problem with the first batch of samples? One of you must have switched them somehow, or damaged them."

He was making this up as he went along; he doubted that there were really any spies, and had no idea what had been wrong with the earlier samples. He hoped that his announcement would not accidentally endanger members of the Watch, but he was fed up with being pushed and shoved and dragged, and words were the only weapon he had with which to fight back. The chance of disconcerting the Governor himself, setting him on a search for nonexistent spies, was too good to pass up.

The medical personnel were glancing warily at one another; his bluff had worked to that extent. The guards, however, were not so imaginative, not so easily confused. One grabbed the front of Starkman's flimsy shirt, thrust his face within a dozen centimeters of Starkman's, and said, "I think you'd better shut up, buddy; none of us are spies or traitors like you and your revolutionary friends."

Starkman asked, "Then how do you explain the first batch of samples?"

"I don't explain anything, and I don't think you do, either. Now, shut up!"

Starkman shut up. He had run out of things to say for the moment in any case, and was beginning to feel a renewed weariness, as well as stinging where the needles had gone in. His body's reserves were clearly still depleted from blood loss.

No one else picked up the conversation after he dropped it; an uneasy silence settled over the lab, broken only by the soft clicking of the machines.

Time passed; the group stood and sat around the room, occasionally moving from one spot to another. No one spoke. Starkman, out of boredom, began to rummage through his shoulder bag again, but was stopped by a cautious guard.

"D'you want to search it again?" Starkman asked, breaking the silence.

"No, that's all right, but just leave it alone for now."

"Fine. For how long? What are we waiting for?"

"We're waiting for the machine's analysis of the cell samples we gave it," replied the elderly doctor.

"Is that the same machine that did the first batch?" Starkman asked.

"Yes."

"Maybe it's the machine that's wrong; maybe it's been sabotaged."

Again Starkman noticed the medical personnel glancing at one another, but nothing further came of it, and he lapsed into silence again.

A moment later a computer's voice spoke from somewhere.

"Bring Mr. Starkman to the Governor's office."

Chapter Fourteen

They took away his pack, but not before he'd managed to palm the flashbomb. He had no idea what was going on, but he thought it might be useful. He tucked it into the waistband of his pants as he stood in the office with his thumbs hooked into his belt-loops.

It was really a very impressive office; one wall was glass, with a view of Capital that made it obvious they were several dozen stories up. Another wall was a copper-tinted mirror, and the remaining two were fine wood paneling. There were four comfortable armchairs, upholstered in red vinyl that did not succeed in looking like leather, and three small wooden tables, and a large potted plant occupied each corner. There was no sign of the Governor, nor of a desk or any other workplace furniture.

The six guards shoved him to the center of the chamber, released him, and left, slamming the door emphatically. He looked about, but remained standing where he was.

"Have a seat, Mr. Starkman," said a voder concealed somewhere.

"Where are you?" he demanded.

"It doesn't matter where I am. Please sit down and make yourself comfortable."

"It matters to me, monster. How do I know I'm talking to a person, and not a machine? How do I know you're anywhere near this room?"

"Why does it matter? All that matters is that we can speak privately and in comfort."

"It matters to me. I want to see you. How do I know

you're really the Governor? Maybe you're just some cheap human thug who's taken over without anyone realizing it. Maybe this whole thing is a fraud contrived by one of the gimcrack dictators from the nineties."

"I am really the Governor, Mr. Starkman, and I am really an alien on your little planet."

"Prove it, then. I've been told over and over that no one on Earth has ever seen you. Why not? I think you're a fake."

"No, Mr. Starkman, I am not a fake. I am sitting less than ten meters from you at this very moment, watching you through one-way glass. It is true that I have not permitted myself to be seen by more than a handful of your species; this is because of the strong tendency toward xenophobia that has been manifest throughout human history. I am very alien indeed, a thing fit to appear in one of your horror movies. Even those who pay lip service to a universal brotherhood of intelligent species seem to assume that I must in some way resemble them; I have read in your fiction and even in your serious speculation that any intelligent being must have hands and legs and eyes. I do not. My chemistry is not based on DNA, but on a wholly different molecular chain—though it is still carbon-centered. My senses are not yours; I said that I was watching you, but 'perceiving' would be more accurate. My air is not the same as yours; my compartment here is a sealed, self-contained environment, with its own atmosphere, temperature, pressure, radiant energy, and biological systems. Only the gravity remains constant, and that is sufficiently weaker than what I am accustomed to that I have trouble in digesting."

"That's a wonderful story, and told in amazingly good English for an alien. You still haven't proved anything. If you're behind that coppered glass, then turn on a light in there and let me see you."

"You are being unreasonable. I brought you here so that I might see you directly and ask you a few polite questions in private before releasing you, and you seem determined to antagonize me."

"I want to see you."

"Mr. Starkman, you said that you had spoken with Imperial spies among my crews."

"Turn on the light and let me see you; I won't answer questions until you do."

"I could force you to answer."

"You could, but I don't think you will. It's much easier to show yourself. Unless you're a fraud, anyway."

"Very well, Mr. Starkman."

The ceiling lights in the office faded to a dull orange glow, and bright yellow light sprang up behind the mirror wall and poured forth. Starkman peered through it at a murky, seething yellowness. He could make out a vague dark mass in the swirling vapor.

"I am afraid that my atmosphere is opaque to your vision. I will approach the glass, so that you may see me, but to do so I must move away from the input for my translator." There was a click, and as Starkman watched, the thing in the other chamber came closer, the golden haze parting before it.

It was an ugly brown color, with glints of shiny metal here and there. Hundreds of stubby protrusions, rather like a protozoan's cilia, writhed on its surface. There were shadowy hollows and odd bumps, but nothing that Starkman could identify with any assurance as eyes or nostrils or any other earthly feature. It had no legs, but was propelled along the floor by the ciliate members; there were no arms or tentacles, but clusters of slightly thinner and longer cilia clutched small unidentifiable devices, like clumps of seaweed tangled around sunken fragments. Metal sparkled in the murk; besides the things it held, parts of its body appeared to be metallic.

It seethed up against the glass, then turned and began retreating once more. As it turned, Starkman saw that a large chunk of its substance had been replaced with a mechanical substitute; metal and something that might have been plastic made up about a fifth of the surface of its back.

He told himself not to jump to conclusions; perhaps that was just clothing, or some sort of armor.

The yellow fog closed around it again, and a few seconds later the voice came from the hidden speakers once more.

"Well, Mr. Starkman, are you now satisfied that I am not human?"

"Not entirely," he lied. "A good Hollywood special-effects crew might have managed it."

"You are a very difficult entity to convince. Could such a crew have imitated the motion of my propelling organs?"

"Maybe not," Starkman admitted.

"May we proceed on the assumption that I am what I claim to be? Human, machine, or alien, I am in power here on your world, and is that not what matters?"

"I guess so. I noticed metal in there; is that part of you?"

"In a fashion. I am a very old being, Mr. Starkman, and over the centuries much of my original body has been cyborged."

"Oh." That, he had to admit, made sense. He had already noticed and commented on the fact that the Governor spoke much more clearly and elegantly through the translator than had Watcher One, or the taped messages from the castaway; he wondered if it might have something to do with the Governor's artificial components.

The lights behind the glass went out, and the office lights brightened; the wall was a coppery mirror once again, the yellow fog and the Governor hidden.

"Now, Mr. Starkman, you claim to have spoken with Imperial spies," the Governor said.

"I lied," he said with a smile as he settled into the nearest chair.

"I would like to believe that, but then there is the question of how you came to know so much about our operations here on Earth, our reasons for being here. How did you learn why we sought you?"

"Why do you care? You have what you came for."

"Mr. Starkman, if I had what I came to Earth for, I would have had you killed by now, and your body vaporized, so that no one else might obtain the same information."

"It wouldn't do you any good; I already gave the information to the other side. There's no point in killing me or destroying Earth to hide it; that would just antagonize people. If you kill me, Earth will rise against you; if you destroy Earth you'll draw the Empire's attention."

"Calm yourself, Mr. Starkman. I hardly think that your death will cause the placid majority of your species to rise up against its benefactor."

"You underestimate our pride—and our xenophobia. The people already resent you aliens coming in here and taking over, and when they learn that you caused the ice age, they *will* revolt."

"They very well might, but who, Mr. Starkman, is going to tell them? You will, not, if we kill you—and if we allow you to live, we will probably find it necessary to erase much of your memory, including your knowledge of the origins of the altered climate. Furthermore, should we let you go with that particular fact still in your possession, and should you then try to broadcast it, we would correct our omission and kill you, albeit belatedly."

"I've *already* told people, monster; several of them, including the members of the Underground."

"And you say you have also given the Empire the information we seek. Tell me, Mr. Starkman, what is this information and how did you give it to them?"

"It's the formula for triggering supernovae, programmed into my DNA and linked to the gene for yellow eyes. I gave them a sample of my blood—how do you think I got this cut on my arm?" He held up the injured member.

"I was aware that you bore a self-inflicted wound; it had not occurred to me that it was done for the purpose of drawing blood, since a simple pin-prick would have sufficed."

"I was in a hurry. I wanted to make sure that they had enough before your goons came and got me."

"So you left this supply of your blood with...with whom, Mr. Starkman? Who are these Imperial spies who have the means of analyzing your genetic material and transmitting the data through interstellar space?"

"I'm not going to tell you that."

"We know who one of them is, of course, and Dr. Curtis is hardly the great spy and revolutionary she would like to be. I doubt very much that she and her comrades would be able to make use of your blood, even had they actually obtained a sample—which they did not, since we took the precaution of taking her into custody as soon as possible after tracing your careless comnet conversation. We had hardly been so gullible as to have believed her protestations of innocence after the incident at the processing center; we released her in hopes that she might lead us to you, as in fact she did. Her every move, every use of the comnet, was carefully recorded. Now, since you took the trouble, knowing that the line might be monitored, of describing our deceased comrade's stay on your world and his ingenious method of concealing information, to the good doctor, I can only assume that it was not from the Underground that you obtained your story of why I am on Earth. This leads me to conclude that you have indeed been in contact with an Imperial spy. Who, Mr. Starkman?"

"I told you, I'm not going to say. Besides, I don't know her name." He thought himself very clever, using the wrong pronoun, until it occurred to him that it might not translate.

"I think that you are intentionally delaying, Mr. Starkman, in hopes of allowing the spy more time to obtain and transmit the data. However, I'm afraid that you're in for something of a disappointment. There is no data to transmit. I told you that you are alive only because I do not have what I came for. I did not lie."

"Why? Why haven't you got it?"

"Mr. Starkman, what color were your father's eyes?"

"Green."

"And your mother's?"

"Brown."

"The carrier that my deceased comrade created had yellow eyes, like your own—and the trait was a simple dominant. Do you know anything of genetics?"

"A little. You mean that I'm not descended from him after all, or one of my parents would have had yellow eyes."

"Exactly. A simple dominant must appear in every generation in order to be passed on; only recessive or partial recessive traits can skip a generation or vanish until reinforced."

"Then I'm just a freak, a mutant."

"This is apparently the case, yes."

"That damn carrier probably died, then, or his descendants all did. The trait could have died out a thousand times over by now."

"This is quite true. It is regrettable that our lost genetic engineer did not think to make his creation longer-lived; then the odds of the trait's extinction would have been less. As it is, we must simply go on looking, on the chance that there are still carriers to be found somewhere on your world."

"I can't stop you from looking, but I think you're wasting your time."

"The matter no longer concerns you."

"I guess it doesn't. Hey, that means that this whole thing has been for nothing; I could have let you catch me any time and no harm would have been done. I've got nothing you want."

"That is not quite true. You had nothing I wanted when you were living in the wastes of your homeland, but you do now. You know something about Imperial spies that you say have infiltrated my organization."

"Oh." That gave Starkman pause. He could not tell the Governor anything about Imperial spies, because he knew nothing. He could, however, tell the whole truth and reveal the existence of the Watch. Compromising his plan for transmitting the data was no longer an issue, since the information apparently no longer existed; the only issue was whether he was willing to betray the organization that had helped him, that was dedicated to peace and non-violence.

He stalled for time. "What happens if I tell you?"

"That depends on what it is you tell me."

"Will you let me live?" If the alien said no, then there would be no point in betrayal.

"That depends on your behavior, and on the reliability of your information."

"You said you might erase part of my memory; can you really do that?"

"Have you never read the many accounts of people

who were kidnapped by our scouts in the middle part of the twentieth century? Many recalled nothing of the events until hypnotized later. We were still unfamiliar with human physiology at the time; now, not even hypnosis will recover what we remove."

If he were to tell the truth, then, he would probably not have to live with the memory and the guilt that would follow. Still, he could not do it. "May I take some time to think about it?" he asked.

"I would prefer to have the information immediately."

"Just a few moments."

"As you wish." The speaker fell silent.

Starkman stood and paced about the room. He paused to lean against the mirrored wall; it was cold, very cold, to his touch, and he recalled that the Governor had said that his compartment had regulated temperature and pressure as well as its own atmosphere.

He wondered how long the alien could live in a terrestrial environment; not long, he was certain. He fumbled with his belt as he considered.

"Mr. Starkman, I must ask for your decision."

He looked up, staring at his own image in the orange glass. "I've decided. I'm not going to betray friends. Take me away and kill me."

"I regret to hear this. You will not be killed immediately; rather, we must resort to less pleasant means of interrogation. Remember that this is of your own choosing."

"I'll remember, monster," Starkman said. He stood and waited, fingers twitching nervously.

A few seconds later the door of the office opened and his squadron of guards reappeared. He fiddled with something at his waist and shifted his weight, letting a small object slide down his leg and fall on the floor by his foot. He tapped it over against the mirrored wall with his toe, hoping that no one saw.

No one did; two of the guards grabbed his arms and headed out the door. Two more led the way, and the final pair lagged behind.

The flash of the bomb was sufficiently startling that

his captors released their hold for a second; that was sufficient. He shoved past the front pair and ran.

As the elevator doors closed in front of him he caught a final glimpse of the six men surrounded by a billowing yellow cloud; they were choking, falling on the carpeted floor of the corridor. He smelled something vaguely unpleasant, and wondered whether he would make it out of the building before the alarms went off. With every passing second, as the elevator descended past floor after floor, he waited for the bells or sirens to sound.

They never did. To his astonishment, he emerged into a ground-floor corridor unhindered, amid people who were obviously unaware of anything amiss.

Though he had difficulty accepting his good fortune, he made use of it, marching on through the mazes of the building, looking for an exit to the street. A purposeful expression and a brisk stride kept the few people he encountered from paying much attention; he relied on speed and keeping his head in motion, back and forth and up and down, to prevent anyone from noticing his peculiar eyes.

By following signs backward he eventually reached a crowded lobby, but hesitated before crossing to the doors and risking the open streets. Cameras were scanning the entryway, he noticed—probably checking for registration numbers.

No, he corrected himself, that wouldn't work; some people kept their hands in their pockets. He couldn't be sure what the cameras were watching.

He turned slightly and stopped at a convenient counter, where he picked up a pamphlet and stared at it while he thought.

He didn't understand why the alarms hadn't gone off. Surely, whether anybody else still alive and conscious had noticed anything or not, Max would know that something was seriously wrong. The Governor's environment must have been computer-regulated; would the alien have used an independent system when Max was on hand? Would such a system have been knocked out by the explosion? That didn't seem reasonable.

He was still surrounded by enemies, even with the Governor dead. He had cut the head off the alien government, but the body would still thrash about for quite some time, he was certain.

Perhaps the alarms were silent to prevent panic. In that case someone or something would be hunting for the saboteur—for him—right now.

He stared at the little brochure, something about services available for recently displaced persons, and tried to think where he could go, what he could do, how he could hide his telltale eyes.

Someone at his shoulder cleared his throat. Starkman glanced up, startled, and saw the Watchman standing at his side, smiling.

"I can't say I approve of your methods," the Watchman said. "But they do get results."

"Are you sure?" Starkman gestured at the milling crowd. "No one seems to have noticed."

"That's our doing." The smile broadened. "I shouldn't be so happy, talking to a man who has just violated our basic principles with a device we ourselves gave him, but I can't help it. It's all working out so well!"

"I don't understand," Starkman said. "*Why* is it working out?"

"I thought you would guess. You did know, after all, that we had partial control of the comnet."

"Max?"

"Yes, Max."

Starkman glanced around. "That's why there were no alarms, then; they're computer-controlled."

"That's right; we had Max programmed to consult us before sounding any sort of alert at all. Up until now we had always passed them on, to avoid any chance of discovery of our tampering, but this was clearly a special case. By the time we found out just what you had done it was too late to save the Governor or those six men—the atmospheres are mutually toxic—but we didn't see any point in letting anyone know about it. In fact, I don't see any reason we should *ever* tell anyone that the Governor is dead."

Realization seeped in as Starkman stared at the Watchman's smile. "That's right," he said. "Nobody ever actually saw the Governor."

"That's right, nobody ever saw it," the other agreed.

"And nobody *will* see it. Max has sealed off that floor for us, and the Governor will go on running the planet for the time being—through Max, of course. The only difference is that Watcher One will be giving Max his orders, rather than the Governor you killed. We didn't dare interfere very much before, because the Governor would have noticed, but nobody will ever question the Governor's own orders. I think you can count on an end to the manhunt—and, as soon as we can manage it, an end to alien government and Earth's independence."

Starkman hesitated before saying, "Won't that make the other Rebels suspicious?"

"They won't argue with their commander," the Watchman said confidently. "The others are all just flunkies; the Governor ran the show."

Starkman was less certain of that, but held his peace. Earth had a chance; he had done that much.

Having done that much, however, led to another question. "What do I do now?" he wondered aloud.

"Whatever you want," the Watchman replied. "Can I give you a ride somewhere? My aircar's outside, and Max has orders to ignore whatever we do and keep the clones out of our way."

Starkman started to say that he had nowhere to go, but then stopped as he remembered Jenny's face.

"Thanks," he said. "I could use a lift to New Denver."